THE PHILOSOPHY OF NIELS BOHR

The Framework of Complementarity

Henry J. Folse
Department of Philosophy
Loyola University – New Orleans

1985

NORTH-HOLLAND
AMSTERDAM · OXFORD · NEW YORK · TOKYO

ISBN: 0 444 86914 x (hardbound edition)
0 444 86938 7 (paperback edition)
1st reprint (Paperback) 1988

Published by:

North-Holland Physics Publishing

a division of

Elsevier Science Publishers B.V.
P.O. Box 103
1000 AC Amsterdam
The Netherlands

Sole distributors for the U.S.A. and Canada:

Elsevier Science Publishing Company, Inc.
52 Vanderbilt Avenue
New York, N.Y. 10017
U.S.A.

Library of Congress Cataloging in Publication Data

Folse, Henry J., 1945
The philosophy of Niels Bohr

Includes index.
1. Complementarity (Physics) 2. Empiricism.
3. Metaphysics. 4. Bohr, Niels Henrik David, 1885–1962. I. Title.
QC174.17.C3F65 1985 501 85–3006
ISBN 0-444-86914-X (hardbound)
0-444-86938-7 (paperback)

Printed in The Netherlands

For Joan

ACKNOWLEDGEMENTS

Research for this book was made possible by a grant from the National Endowment for the Humanities, Fellowships for Younger Humanists Program. I would like to express my appreciation to the people of the Niels Bohr Institute who made our stay in Copenhagen so very pleasant, and in particular to Professors Erik Rüdinger and Jørgen Kalckar who read my manuscript and made invaluable suggestions, as well as to Professor Frank Durham of Tulane University. Work on this manuscript was also made possible by grants from The College of Charleston, Charleston, South Carolina, and Loyola University, New Orleans, Louisiana.

TABLE OF CONTENTS

INTRODUCTION

Certainly of all the developments in twentieth century physics, none has given rise to more heated debates than the changes in our understanding of science precipitated by the "quantum revolution". In this revolution Niels Bohr's dramatically non-classical theory of the atom proved to be the springboard from which the new atomic physics drew its momentum. Furthermore, Bohr's contribution was crucial not only because his interpretation of quantum mechanics became the most widely accepted view but also because in his role as educator and spokesman for atomic physics Bohr was very much the patron spirit of the entire quantum revolution. The conceptual framework which he proposed to provide a new viewpoint for understanding the quantum theoretical description of atomic systems became for most of this century the dominant outlook of countless productive experimental and theoretical physicists. He called this new framework "complementarity".

Thus it comes as no surprise that virtually all discussions of complementarity have approached the subject primarily from the point of view of the correct interpretation of quantum physics. This situation is somewhat unfortunate, for Bohr not only saw complementarity as a new framework which would resolve quantum paradoxes, but also regarded it as teaching an "epistemological lesson" pertinent to all description of nature. Quite naturally those who have been interested in the correct interpretation of quantum mechanics have for the most part been conversant with the mathematical formalism of quantum theory. Consequently, discussion of complementarity has tended to be restricted to persons with expertise in physics or philosophy of physics.

Originally this work was intended to be a treatise on the philosophical significance of complementarity. Yet in spite of the immense importance of Bohr's viewpoint both as regards the history of twentieth century physics and its philosophical impact, misunderstandings and confusions about complementarity abound. Thus in the course of my work, it emerged that what was needed foremost was a detailed analysis of exactly what Bohr did and did not assert about the description of nature in the atomic domain. In particular it seemed that the greatest good could be accomplished by presenting Bohr's framework of complementarity in a way which did not require the expertise in physics and mathematics of necessity presumed by most discussions of the interpretation of quantum theory.

Consequently, here I propose to bring complementarity to a much broader audience, including all who can appreciate Bohr's ideas in the wider sense of expressing a way of understanding how the concepts employed by science in its description of nature relate to the physical world they are used to describe. Indeed, I would hope to find among my audience all who accept the belief that physical science has relevance for our philosophical understanding of the nature of reality. Fortunately, the nature of Bohr's argumentation is such that he never required any advanced knowledge of physics to understand his central message. Thus it is possible to present complementarity in its historical genesis and development in a way accessible to the general reader for whom the developments of twentieth century physics are of interest from either a philosophical, an historical, or a scientific point of view. This book is intended to provide just such an analysis.

Anyone familiar with the literature on quantum theory knows all too well that Bohr's ideas have been the subject of many an interpretation and that there are numerous "quantum philosophies", some supporting Bohr and others stridently critical. Since I do want very carefully to guard against partisan misinterpretations of Bohr's views, it is difficult to present my interpretation of complementarity without appearing to offer a polemical defense of its particular account of the paradoxes of quantum physics. But my goal is understanding Bohr's philosophy, not resolving the quantum mystery or showing that Bohr really belongs to this or that school of philosophy.

Consequently, the nature of the task before us is such that it is difficult, if not impossible, to present complementarity without at the same time "interpreting" it. This is, then, an interpretation of Bohr's philosophy which attempts to present it as a coherent framework for the description of nature. Therefore I have shunned the posture of a hostile critic bent on pointing up difficulties in Bohr's arguments or inconsistencies in his comments. Instead, I have adopted a manner of exposition which reconstructs Bohr's viewpoint in what appears to me to be its best possible form. Furthermore, this interpretation appears to be in the best possible accord with what I intend to show to be the historical path that Bohr actually followed to complementarity. However, the fact that there are other interpretations, or, if you will, misinterpretations, of complementarity implies that while presenting Bohr's viewpoint, I shall also have to argue against what I perceive to be the most important misunderstandings about complementarity.

These misinterpretations seem to result from two factors. In the first place virtually all analyses of complementarity have taken place within the context of discussions on the correct interpretation of quantum theory. Without intending to disparage the crucial importance of that question, it is also clear that as long as it is the central one of any study, the correct interpretation of Bohr's philosophy will always be something of a secondary objective. Just how secondary is often rather appalling. Thus it is not uncommon to find discussions of the interpretation of quantum theory which characterize Bohr's complex and subtle point of view in a sentence or two, without more than a passing quotation from its supposed author. Even when somewhat more care is given to understanding complementarity, as often as not, that attention is bestowed only for the sake of arguing for or against some particular interpretation of the theory which may very well be quite distinct from, if not opposed to, Bohr's interpretation. In the second place, such treatments of Bohr's viewpoint generally proceed from the finished version of quantum mechanics, often in a mathematical formulation which appeared only well after Bohr had developed his new framework. In fact Bohr never tried to reduce his viewpoint to a few basic principles, nor did he ever use any sophisticated formalized presentation of the theory as the basis of his arguments. Such ahistorical approaches hardly seem well suited to understanding complementarity as Bohr must have understood it.

Thus in presenting my analysis and interpretation of complementarity I have adopted three guiding methodological assumptions. *First* the general question of the interpretation of quantum mechanics (which I regard to be the proper concern of specialists in the foundations of physics) is to be set to one side in favor of the direct goal of understanding Bohr's philosophy and how he arrived at it. Obviously since complementarity is designed to interpret quantum theory, in arguing for what I take to be the correct understanding of complementarity, I am in effect defending the virtues of a particular interpretation of quantum theory, and in general of the significance of the whole quantum revolution. However, in order not to obscure the understanding of complementarity which is my central focus, in defending the interpretation of quantum theory provided by complementarity, I have avoided all reference to other, rival attempts at an interpretation of this controversial theory. Thus this is not a work on the various interpretations of quantum theory and/or how they relate to Bohr's.

Since my direct concern is with Bohr's philosophy of complementarity, not the interpretation of quantum mechanics, here it is possible to separate the empirical theory of quantum mechanics from its formal expression in the mathematical formalism. Thus in the following chapters I discuss the description of atomic systems made possible by the quantum revolution in a purely conceptual, non-mathematical fashion. No doubt this approach may be welcome to those readers not trained in mathematical physics, but it may give pause to those readers who are able to follow the theoretical formalism. Were the issue at hand the correct interpretation of quantum theory, then this manner of presentation would not be possible. However, since understanding Bohr's philosophy allows us to put this issue to one side, so to speak, it seems to me justified to introduce a far larger audience to Bohr's thought by omitting all reference to the mathematics.

Although this approach may seem suspect to some readers, it should suffice to put such suspicions to rest by reminding them that I am doing no more than following Bohr's lead in this respect. He never presented his arguments as derived from any source other than a purely conceptual analysis of the requirements for an adequate description of atomic systems. To those readers familiar with the mathematics of quantum theory, I must beg pardon for sacrificing the technical rigor of a formal account. However, this approach should not be seen as merely a concession necessary to reach a broader audience. I hope that it will become clear in the following chapters that such a move is positively conducive to understanding what Bohr claimed in complementarity and how he reached these conclusions. Indeed, I would maintain that by clearing the field of technical questions which are the proper domain of the specialist in atomic physics, complementarity will not only be understandable to a far broader audience, but also will be seen for its full philosophical significance.

Of course I must follow how Bohr's arguments are intended to establish how the quantum revolution forces a generalization of the classical framework, but my presentation will differ from his in method as it differs in goal. The typical detailed discussions of various thought-experiments, at which Bohr was an incomparable master, were intended to show how complementarity resolves quantum paradoxes. But since this is incidental to our central concern and since such discussions tend to act as a barrier to the reader untrained in mathematical physics, the analysis of thought-experiments which figures so prominently in some of Bohr's presentations will assume a very secondary role in my treatment. However, the epistemological and ontological discussions which are so frustratingly brief in Bohr's essays will be correspondingly magnified in mine.

Secondly, since it appears that many misconceptions about complementarity have resulted from the failure to understand complementarity in its historical genesis, I have made my approach to complementarity follow the historical path that Bohr followed to his new viewpoint, as best it can be reconstructed. Thus this work includes a history of the origin and development of complementarity, starting in the quantum revolution and extending to its epistemological lesson for other fields of science as well as for philosophy of science. But its historical content is solely for the purpose of understanding complementarity, not for presenting the story of the quantum revolution or the influences on the development of atomic physics that did not affect directly Bohr's formulation of complementarity. Nor should my analysis be viewed as an intellectual biography of Bohr as a natural philosopher, although the development of his ideas is so closely bound up with his intellectual life that it is difficult to write on the subject without to some extent touching on his biography.

Since the time of Socrates it has been a favorite tack of philosophers to take a specific issue and follow the widening circles of intellectual change necessitated by the analysis of an apparently restricted problem, much like ripples spreading from a pebble dropped in the still waters of the prephilosophical outlook. So it is my intention to trace how Bohr's reflections on a specific problem in physics, introduced by theoretical demands of limited scope, spread to ever widening themes, leading ultimately to a profound and suggestive revision of our understanding of the concepts on which we rely for the scientific description of nature. Therefore, although my *prima facie* goal is expository, this analysis contains a lesson within philosophy itself. The issue here is not simply how to understand Bohr's thought; it is an attempt to come to grips with the fundamental question of the relationship between empirical knowledge gained in the physical sciences and those epistemological and ontological presuppositions which are necessary to justify our belief that such knowledge provides an objective description of the physical world. Nevertheless, I have been careful to write in a language which does not exclude those not conversant with the professional vocabulary of philosophers of science.

On the one hand it is indeed essential to appreciate the fact that complementarity had its birth in the quantum revolution, but its philosophical maturity will be attained only when it is understood as enunciating a general framework for the description of nature in all empirical science. On the other hand, although Bohr insisted that complementarity taught an epistemological lesson which extended beyond the borders of quantum theory, it would be perverse and distort the actual path of Bohr's thinking to present complementarity abstracted from the concrete problem of formulating a physical description of the atom. Bohr's overpowering drive to penetrate the secrets of the atomic system forced him to come to grips with the problem of how a conceptual description of nature is achieved. His concern with the philosophical basis of scientific description went hand in hand with his long efforts to provide a secure defense for his conviction that by 1927 quantum theory had achieved in a certain sense a *complete* description of the physical system of the atom. In the following chapters we shall see that Bohr never separated these two aspects of his thought.

While Bohr himself understood from the beginning that he was concerned with philosophical issues extending far beyond his proposed solution to the specific quantum paradoxes that have held the center of attention since 1927, unfortunately, history has not been altogether kind to his philosophical endeavors. Instead of being understood as a general framework within which the new physics was to be justified as an objective description of nature, complementarity came to be identified with the so-called "Copenhagen Interpretation" of quantum theory. Furthermore, this Copenhagen Interpretation, and complementarity with it, came to be commonly associated with the writings of a whole group of physicists, who may not have always fully grasped what Bohr was saying. Thus proclamations about complementarity and quantum physics made under the banner of the "Copenhagen Interpretation" are generally imputed to Bohr, though in fact Bohr himself almost certainly never intended such views. Hence in analyzing complementarity I have adopted a *third* methodological assumption of relying only on Bohr's writings. Therefore I have neglected all comments made by way of defense or exposition by Bohr's followers. I have, however, relied heavily on the testimony of these physicists for historical points or matters relating to Bohr's methods or general intellectual outlook.

Though Bohr recognized that the "epistemological lesson" which quantum theory had taught was applicable to fields other than atomic physics, unfortunately his attempts to draw the consequences of this lesson into other fields generally appear in his essays only after considerable discussion of atomic physics. Thus they have the appearance of merely a speculative epilogue. If we add to this fact the difficulties that Bohr's total opus has no systematic structure, but covers the same ground over and over again in brief essays written to be presented as public lectures, that such essays lack the detailed exposition a book-length study could have afforded, and that Bohr's writing style is densely packed and rarely transparent, then we have ample case to justify reconstructing Bohr's philosophy as a whole.

But for these same reasons, attempting to reconstruct complementarity as a general conceptual framework inevitably requires considerable latitude for interpretation of Bohr's writings. Given the nature of these writings, it seems quite out of the question to achieve a coherent formulation of complementarity as a framework for describing nature without a considerable interpretive addition to what Bohr explicitly states. Nevertheless, if we keep Bohr's own goals foremost in our minds (instead of allowing the goals of other philosophical positions to take precedence), and if we follow carefully the route that Bohr's thought actually took as he developed his philosophy, then we need not be unduly pessimistic about the degree to which such an interpretation may accord with Bohr's intentions.

While there is ample justification for presenting Bohr's ideas as extricated from the context of quantum problems, here I have also found it occasionally necessary to interpolate more than what Bohr explicitly states. Though part of the fascination with Bohr as a thinker results from his incessant concern with questions beyond the customary domain of physics, he was a sufficiently scrupulous thinker—as well as being sensitive to the weight his reputation gave his every utterance—not pontificate on subjects beyond his expertise. Thus while his concerns are obviously philosophical, because he was not expert in the history of philosophy or the philosophy of science, he carefully refrains from commenting on the relations of his own thought to the traditional questions and positions of philosophy.

Since I am writing for philosophers, historians, and scientists whom I presume to have an interest in Bohr as a natural philosopher, it is inevitable that I refer his thought to a central issue in the philosophical analysis of scientific knowledge, that which separates the defenders of scientific "realism" from its critics. In particular I propose to argue that Bohr is properly understood as defending *a* form of realism, *i.e.* as upholding the view that it is the task of science to understand the nature of physical reality as it exists independently of the phenomena through which we experience it. However, so as not to lose those readers whose primary interest is history or simply what Bohr said about quantum paradoxes, I have reserved discussion of the relations between complementarity and specifically philosophical issues for the last two chapters. In the earlier chapters, the only philosophical issue before us is understanding Bohr's philosophy. Furthermore, in order not to lose those readers who are untrained in the philosophy of science, even in the last chapters I have taken care to define in context those purely philosophical terms which it is necessary to introduce.

One cannot be *certain* that Bohr would have greeted all of these arguments in the last two chapters with unqualified approval, for here I have obviously gone beyond what could be considered Bohr's direct contribution to complementarity. Though the failure to understand complementarity on the part of philosophers certainly disillusioned Bohr about the significance of most traditional or then contemporary philosophical approaches to science, we must recognize that the misunderstanding was mutual, not one-sided. As we shall see, Bohr's lack of familiarity with philosophical terminology and positions ultimately hindered his ability to communicate the philosophical significance of his new framework. Nevertheless, concerns with the unity of knowledge and the structure of experience as they affect the use of concepts are certainly Bohr's overriding concerns as well as central themes in philosophy. In sharing with him a common regard for the broad perspective of scientific knowledge, I hope that this commentary on his thought will be in greater sympathy than those treatments which discuss complementarity only in relation to the correct interpretation of quantum theory. If this effort succeeds in bringing complementarity into a wider arena, it will have achieved its function.

CHAPTER ONE

WHAT IS COMPLEMENTARITY?

Late in his life Niels Bohr remarked, "I think that it would be reasonable to say that no man who is called a philosopher really understands what is meant by the complementary descriptions".[1] That somewhat wistful comment by this great pioneer of modern atomic theory is as sadly true today as it was over fifty years ago when Bohr first formulated the philosophy he called complementarity. This book aspires to remedy that situation by presenting Bohr's philosophy as a consistent and comprehensive framework for the objective description of nature. Before beginning our analysis of complementarity, it will aid our understanding to recognize what is involved in identifying its genus as a "conceptual framework". Thus in the first section of this opening chapter I will present what Bohr meant by proposing complementarity as a conceptual framework for the description of nature, but I must do this before having the chance to develop the arguments by which he arrived at this framework. Hopefully whatever loss in dramatic presentation results from this foreknowledge of our goal will be offset by a greater ease in following the arguments in chapters to come. In order to clear the way for our further understanding of complementarity, in the following sections I shall consider several common misconceptions about complementarity and how they arose and then turn to the significance of complementarity as a framework.

1. Complementarity as a Framework for the Description of Nature

With characteristic understatement Bohr generally chose to refer to complementarity as a "general viewpoint" from which the paradoxes of quantum theory could be resolved.[2] He arrived at this new "viewpoint" by arguing that the quantum revolution had made unavoidable a fundamental revision of the conceptual basis on which the scientific description of nature rests. By adopting the viewpoint of this framework, Bohr argued one could eliminate the paradoxes which arise when the physical description provided by quantum mechanics is viewed from the framework of classical physics. Thus the apparent modesty of the phrase "general viewpoint" is rescinded when he reveals that "the essential inadequacy of the customary viewpoint of natural philosophy for a rational account of physical phenomena ... entails ... the necessity of a final renunciation of the classical ideal of causality and a radical revision of our attitude towards the problem of physical reality".[3] Here is a clearly revolutionary call for a fundamental change in our philosophical understanding of the scientific description of nature.

Thus, following Bohr, I will treat complementarity as a "general viewpoint" from within which the description of physical phenomena in natural philosophy is to be understood. In accepting this viewpoint we should be prepared to revise radically the position of the "customary viewpoint" on the relationship between the description of phenomena in natural philosophy and the nature of physical reality. But Bohr's reference to complementarity as a "general viewpoint" should not mislead us into thinking that it is an *arbitrary* construction of thought. He is always careful to emphasize that we are *forced* to adopt the complementarity viewpoint by the unsuspected empirical discovery that we must adopt what he called "the quantum postulate" in order to develop a successful description of the phenomena of atomic physics. It was the momentous decision to adopt this dramatically non-classical assumption that enabled him to formulate his 1913 model of the atomic system and that brought him into the center of atomic theory in this century. As Bohr understood it, his decision to adopt this postulate is not merely the consequence of a way of viewing nature which we may or may not choose to adopt. Like Newton, he eschews the position that science is to be built upon hypotheses in the rationalist sense:

> We are not dealing with a hypothesis or any arbitrary viewpoint, but we are only using a word to make a short reference to the argumentation which describes our position as observers in such a field of experience where ... we are in the position that the various experimental arrangements under which phenomena can occur exclude one another. ... We call such phenomena "complementary" just to use a word—a fairly short word—so that we won't have to repeat the whole argument all of the time.[4]

Thus complementarity claims to be based on empirically corroborated postulates within which the description of those phenomena which support quantum theory are to be understood. No doubt Bohr failed to define the word "complementarity" as precisely as we might like, but it is clear that he uses it to refer not only to the framework he proposed but also to the arguments designed to establish that framework as a rational generalization of the classical framework:

> The idea of using a word like "complementarity" and its possible uses is namely that it is a new artificial word; a word which cannot call upon any ideas about situations. It's only to remind about such types of arguments as are drawn from atomic theory. The point is that it points to, I'd rather say, a wider frame in which it has been possible to comprehend the laws of nature ... the existence of which are incompatible with the ordinary accustomed principles of natural philosophy.[5]

Unfortunately, the phrase "general viewpoint" does not take us much further in understanding exactly the sort of a thing Bohr intends complementarity to be. However, we can get closer to our goal by considering two of Bohr's favorite points about complementarity. The revolutionary call for a new viewpoint for understanding the description of atomic phenomena should be the most striking face which complementarity presents, but it is not the one Bohr emphasizes in his earlier papers. He was certainly not one to hedge his bets when it came to the need for a radical departure from accustomed conceptions. But he was too driven by a desire to maintain the continuity of physics to speak only of revolutionary changes. Thus he tried to present complementarity as the rational consequence of combining the non-classical "quantum postulate", made inescapable by atomic physics, with the older conceptual framework of classical physics. This is what he has in mind when he speaks of complementarity as a "rational generalization" of the "classical ideal of causality".

When Bohr speaks of complementarity as rational generalization of the classical ideal of causality, it appears as a new framework derived from the empirical demand to accept the "quantum postulate" and yet make atomic physics consistent with the continued use of the classical concepts to describe those phenomena through which we confirm this theoretical account of the structure and behavior of atomic systems. Later we shall see why Bohr regarded the presuppositions of his new framework as "more general" than the specific presuppositions of the classical framework. For the present we need only note that when he speaks of a "rational generalization" he means to refer to how this new framework is derived from its predecessor. As a physicist, Bohr's professional interest was to show how his rational generalization of the classical framework resolves the paradoxes or inconsistencies which appear when quantum theory is understood from within the classical framework. Therefore the central argument of all of Bohr's writings intends to establish from the empirically confirmed quantum postulate, which Bohr simply accepts as a contingent fact of nature, the necessity of such a generalization.

The second favorite point Bohr makes is seen in his frequent references to the "epistemological lesson of complementarity". In using this expression Bohr means to refer to a group of arguments aimed at showing the necessity for revising the "principles of natural philosophy" on which the classical description of physical phenomena rests. The philosophical significance of this claim dawns on us only when we see that Bohr intends his arguments to demonstrate no less than the need for replacing, or at least radically revising the "framework of classical physics" so painstakingly built up over the centuries from Galileo to Einstein. Thus Bohr's epistemological lesson is no by-the-way admonition to pick up a bit of useful information from physics. It is no less than an attempted full-scale revolution in our understanding of how the conceptual framework of science makes possible the description of nature. This point must be emphasized at the outset, for the failure to understand the genus of complementarity has led to much misunderstanding and irrelevant criticism.

Throughout the development of quantum theory Bohr repeatedly stressed that venturing into this new domain of human knowledge requires "a constant extension of the frame of concepts appropriate for the classification of new experiences". In turn, this "leads to a general epistemological attitude which might help us to avoid apparent conceptual difficulties in other fields of science as well".[6] Consequently, it is clear that Bohr intended to establish complementarity as a contribution "to the general philosophical clarification of the presuppositions underlying human knowledge".[7] Thus in no way should complementarity be restricted to the analysis of only the specific paradoxes which arise in the quantum mechanical description of atomic systems. In Chapter Six we shall see how Bohr argued that complementarity could resolve persistent paradoxes in the description of biological and psychological phenomena.

Indeed, if, as Bohr maintains, complementarity is to be considered as a parallel to the "whole mode of description of classical physics",[8] then it cannot be concerned exclusively with the interpretation of quantum physics, but must instead address itself generally to the analysis of the use of concepts in what Bohr called the "description of nature". Thus we may expect that the epistemological lesson of complementarity not only will teach us something about the use of the specific descriptive concepts employed by quantum physics and other specific sciences, but also will have something to say in general about any use of concepts in the description of nature.

In its epistemological lesson complementarity includes an account of how conceptual frameworks are refashioned through the progress of science. The use of any particular framework of concepts to describe certain phenomena entails presupposing that the phenomena in question are of such a nature that they can be described by these concepts. Such presuppositions may be set forth clearly in a particular theory, but as the theory is modified and extended to encompass more phenomena the use of the same descriptive concepts may well be subtly altered such that the original presuppositions become no longer applicable. Thus Bohr observes that the ...

> ... history of science teaches us again and again how the extension of our knowledge may lead to the recognition of relations between formerly unconnected groups of phenomena, the harmonious synthesis of which demands a renewed revision of the presuppositions for the unambiguous application of even our most elementary concepts.[9]

History discloses that unless the presuppositions governing the applicability of concepts to the description of nature are periodically criticized and revised, the expansion of knowledge into new domains will generate paradoxes and inconsistencies. For this reason, Bohr summarizes his epistemological lesson as follows:

> ... Often the development of physics has taught us that a consistent application of even the most elementary concepts indispensable for the description of our daily experience, is based on assumptions initially unnoticed, the explicit consideration of which is, however, essential if we wish to obtain a classification of more extended domains of experience as clear and as free from arbitrariness as possible. ... This development has contributed to the general philosophical clarification of the principles underlying human knowledge. ... We have nevertheless received only recently an incisive admonition that the analysis of new experience is liable to disclose again and again the unrecognized presuppositions for an unambiguous use of our most simple concepts such as space–time description and causality.[10]

The phrase, "the description of nature" provides a key to understanding Bohr's epistemological lesson. It appears in the title of Bohr's first book of philosophical essays, *Atomic Theory and the Description of Nature*, as well as many times in the titles and contexts of essays and lectures. As Bohr understands science, "nature", that of which we are aware in experience, is taken as the given field of phenomena to be described by whichever conceptual framework can be shown to achieve the greatest possible consistent order or "harmony" in establishing the regularities observed among phenomena while also preserving the widest possible scope. Bohr refers to this task as "analysis and synthesis". The concepts of the framework must permit the "analysis" of specific phenomena, for "the very essence of scientific explanation [is] ... the analysis of more complex phenomena into simpler ones".[11] Though he does not explicitly define "conceptual framework", science requires such a conceptual framework to interrelate the descriptions of all experiences: "When speaking of a conceptual framework, we refer merely to the unambiguous logical representation of relations between experiences."[12] Although the framework is fashioned by the scientist, or more accurately by the collective labors of generations of scientists, Bohr holds that it is nature, not the scientist, that determines which framework will permit such descriptions. As the scope of scientific knowledge grows to encompass evermore phenomena or to weld together diverse aspects of scientific knowledge, nature forces revisions or "generalizations" of the older frameworks of science. As Bohr puts it, complementarity provides "a widening of our conceptual framework for the harmonious comprehension of apparently contrasting phenomena".[13] The fact that a conceptual framework is to be judged by the extent to which it makes the description of nature "harmonious" is evident when he praises Newton's achievement on the grounds that "through his conception ... our whole world picture achieved a higher degree of *unity* and *harmony* than ever before".[14] Similarly he claims that complementarity preserves a "harmonious synthesis" in its conceptual framework: "In offering a frame wide enough to allow a harmonious synthesis of the peculiar regularities of atomic phenomena, the conception of complementarity may be regarded as a rational generalization of the very ideal of causality."[15] Thus complementarity is a conceptual framework which seeks to achieve a comprehensive description of nature by establishing a harmonious viewpoint for the analysis and synthesis of phenomena.

The phrase "unity of knowledge" also often appears in Bohr's terminology. It expresses Bohr's strong conviction that all the sciences should provide a single consistent, harmonious description of the one world which is the object of all human experience, of "nature". The scientific description of nature is then inseparably linked to the philosophical task of formulating a comprehensive framework for a description of all nature. But the necessary condition for the use of any framework in science is that it describes nature "objectively". Thus Bohr constantly balances his demand for a new framework with the emphasis that scientific knowledge as such is shaped by this demand for objectivity. While alternative frameworks may offer opportunities for the description of experience in private, subjective—and therefore "ambiguous"—terms, the framework of science is essentially determined by the objectivity and unity towards which its description of nature must aim.

While Bohr is adamant that complementarity provides an objective description of phenomena, in discarding the classical framework, he discards with it its criteria for objective description. Thus complementarity involves a new account of objectivity which we shall examine in Chapter Seven. For Bohr the problem of objectivity does not refer to how subjective experiences can represent an "external world" (with the consequent attempt to base objectivity on a foundation lying outside of experience) but instead refers to the task of formulating a framework adequate for the *unambiguous* description of nature. The epistemological lesson of complementarity thus requires a "new background" for the use of words like "objectivity":

> Common to the schools of so-called empirical and critical philosophy, an attitude therefore prevailed of a more or less vague distinction between objective knowledge and subjective belief. By the lesson regarding our position as observers of nature, which the development of physical science in the present century has given us, a new background has, however, been created just for the use of such words as objectivity and subjectivity. From a logical standpoint, we can by an objective description only understand a communication of experience which does not admit of ambiguity as regards the perception of such communications.[16]

From this new perspective the central function of the epistemological analysis of objective knowledge must be the determination of a framework through which unambiguous communication can be achieved. But as the expansion of knowledge into new domains of experience brings about the often tacit casting aside of older presuppositions, a description which was once perfectly unambiguous may no longer be so, at least not until the framework itself is revised. Thus Bohr observes:

> Every scientist, however, is constantly confronted with the problem of objective description of experience, by which we mean unambiguous communication. ... The main point to realize is that all knowledge presents itself within a conceptual framework adapted to account for previous experience and that any frame may prove too narrow to comprehend new experiences. Scientific research in many domains of knowledge has indeed time and again proved the necessity of abandoning or remoulding points of view which, because of their fruitfulness and apparently unrestricted applicability, were regarded as indispensable for rational explanation.[17]

Thus it is nature, the object of description in science, which determines the objectivity of a description by forcing science to remold its framework in accord with new presuppositions for unambiguous communication which we learn as our experience of nature expands into new fields.

As a consequence of this view, the "objectivity" of a description becomes the ultimate criterion guiding any revision of a framework undertaken as a result of adopting new presuppositions about nature. Bohr argues that this demand for objectivity, which is satisfied by the necessary and sufficient condition of unambiguous communicability, requires us to analyze how the concepts of our ordinary language developed for describing experiences function to permit unambiguous communication:

> As the goal of science is to augment and order our experience, every analysis of the conditions of human knowledge must rest on consideration of the character and scope of our means of communication. The basis is here, of course, the language developed for our orientation in the surroundings and for the organization of human communities.[18]

Here Bohr claims that the need for the new framework of complementarity results from the necessity for describing phenomena in the unambiguous language adapted for communicating everyday experiences, and this is none other than the language which employs the concepts of classical physics, those of space–time description and causal connection. Thus he is saying both that certain fundamental concepts are unavoidable for describing all phenomena, and at the same time calling for a wholesale revision of the conceptual framework within which such descriptions are expressed.

Though it may seem that such a position is inconsistent, the apparent problem is removed when we recognize the distinction between the empirical reference and the theoretical meaning of a term within a conceptual framework. The ambiguity which an objective description must overcome is not a simple ambiguity of word meaning within a given language. Rather, what Bohr claimed is that as our understanding of nature is broadened by describing new experiences, concepts once adequate for describing nature may no longer have clear meanings unless we take care to criticize and revise the presuppositions governing their use. Such revisions then logically entail altering the conceptual framework within which the concepts are used. At least some of the concepts employed within any theoretical description must have some empirical reference, or in other words refer to some aspect of sensory experience. However, those same concepts are also given a theoretical meaning within the context of a theoretical description of phenomena. Since their theoretical meaning is at least partially determined by the framework of presuppositions stating the conditions governing the applicability of such concepts, a change of framework will alter the presuppositions governing the use of descriptive concepts even while the empirical reference of those concepts remains fixed.[19]

For example, the concept "red" as a particular sensory response refers to an aspect of perceptual experience determined by the visual faculty of the perceiver and made unambiguous by assuming the "intersubjective" constancy of that visual faculty from one normal perceiver to another. This empirical reference is quite independent of conceptual framework. But within the conceptual framework which makes possible the electromagnetic theory of light, the concept "red" attains a distinct theoretical meaning apart from its use to refer to a particular aspect of experienced phenomena. Bohr's point is that when in order to describe certain phenomena, we adopt a theory such as the electromagnetic theory of light, we implicitly accept certain presuppositions for the applicability of color concepts in the descriptions of objects, such that, for example, the "color of an electron" becomes a meaningless notion while the "color of a billiard ball" retains precisely the empirical reference it had prior to the acceptance of the theory. Thus the empirical reference of concepts used in communicating any description of experience may remain constant (though perhaps not definable linguistically) and yet the framework for describing experience may be radically altered by changing the presuppositions governing the applicability of such concepts to various objects. Indeed, such a revision may be entailed by the empirical discovery, should it occur, that any previously accepted presuppositions effectively preclude the possibility of describing an ever growing range of phenomena. Such a discovery is particularly likely to occur when the description of nature enters new domains of phenomena. Thus the scientist who pushes back the frontiers of knowledge should be prepared to modify even the most strongly held of such presuppositions. And in fact this is precisely what Bohr believes has happened in the quantum revolution. Thus his epistemological lesson turns on an argument governing the use of concepts in this latter sense, *i.e.* with respect to the presuppositions involved in determining whether any given descriptive concept is unambiguous in a given context in which it may be employed.

2. Common Misconceptions about Complementarity

Time and again numerous authors writing on the significance of twentieth century physics have claimed the territory of complementarity under a variety of mutually hostile banners. It is appropriate to begin this study by emphasizing that complementarity is a conceptual framework designed to replace that of classical physics because it is precisely the failure to come to grips with this fact that has led to such misunderstandings of Bohr's view. Thus the failure to understand complementarity as a framework has led to the somewhat absurd situation that each of the various schools of philosophy of science of this century has, from one philosopher to another, claimed Bohr as both ally and enemy. He has been championed as a positivist, a realist, a materialist, an idealist, and a pragmatist, *and* at the same time criticized under at least as many labels. One might suppose that this state of affairs indicates confusion on Bohr's part, but though it may partially stem from Bohr's philosophical innocence, the confusion is basically the failure to understand the nature of his philosophical viewpoint.

Detailed arguments intended to reject these claims must wait until later chapters, once we have a clearer idea of the framework and its epistemological lesson. But at this point it will help us to get a feel for complementarity to point out a few of the most common misconceptions about Bohr's viewpoint. The simple fact of the matter is that there is much confusion concerning the relationships between complementarity and quantum theory and between it and the classical framework. Thus the failure to comprehend that complementarity is a framework has led to two sorts of misunderstandings. The first sort arise from the tendency to regard complementarity as a principle or theory *within* quantum physics and the second sort arise from viewing complementarity from the classical framework which it was designed to replace. Adopting either or both of these postures towards complementarity, will result in a failure to appreciate Bohr's arguments.

Although Bohr's pupils and defenders are fond of referring to the "*principle* of complementarity" or the "*theory* of complementarity", in no place in Bohr's writings—including both published essays and private letters and manuscripts—have I ever found any reference to such an alleged "principle" or "theory", much less to an explicit statement of any such principle or theory. In fact, Bohr wisely avoided such terms, for speaking in this way would tend to produce misunderstandings regarding the sort of a thing he intended complementarity to be. "Complementarity" never labels any principle or theory, and searching for such merely obscures the fact that it is a "conceptual framework" from which to view physical princples or theories.

The tendency to treat complementarity as a "principle" within quantum physics has led to its common identification with the Copenhagen Interpretation. It may come as a surprise to many readers that the phrase "Copenhagen Interpretation" was a term Bohr never used and certainly regretted. Although one might quite correctly point out that "complementarity" indeed provides the "general viewpoint" on the basis of which the so-called "Copenhagen Interpretation" presents its justification for accepting quantum theory as a complete description of atomic systems, it is possible to distinguish between the two. While the Copenhagen Interpretation of quantum theory involves lessons drawn from complementarity, it relies additionally on other *physical* principles which need not be taken as part of the framework of complementarity. Thus Bohr would have argued that complementarity could be reached from a logical analysis of the use of concepts once we adopt the nonclassical assumptions forced upon us by the quantum revolution. However, the Copenhagen Interpretation requires additional physical postulates that can be confirmed only by the empirical testing of the theoretical predictions of quantum mechanics. It is because of this tendency to collapse the distinction between complementarity and the Copenhagen Interpretation that the views of other physicists who espouse the Copenhagen Interpretation are often misidentified as Bohr's views. It is no wonder then that such a melange of different, and very likely incompatible positions, is often found to be incoherent by critics of this so-called Copenhagen Interpretation. Here I will equate "complementarity" with only Bohr's view.

From the standpoint of the classical framework, the two aspects of quantum physics which stand out as most paradoxical are Heisenberg's uncertainty principle and the duality of particle and wave "pictures" of atomic phenomena. In Chapter Three we will follow the origin of these discoveries and their impact on the development of Bohr's thinking. But at this time it is appropriate to examine how these discoveries have led to mistaken notions about complementarity.

The "uncertainty" or "indeterminacy" principle was formulated by Werner Heisenberg in 1927 while working in close collaboration with Bohr.[20] This principle played a crucial role in clarifying Bohr's own ideas about complementarity, a role which we will examine in Chapters Three and Four. However, it can be documented that Bohr was already well on his way to the ideas of complementarity well before Heisenberg made his discovery. Thus one cannot assume that Bohr formulated complementarity merely to explain or justify this celebrated departure from the strict determinism which is the ideal of the classical framework. However, Bohr's intention to resolve the quantum paradoxes by adopting a radically new framework led him to expect that the classical goals may have to be abandoned. He saw this possibility as confirmed by the fact that the uncertainty principle implied that the classical ideal of causality, strict determinism, would have to be replaced by a new goal for an adequate description of the behavior of a system, that expressed by statistical determinism.

The relation between complementarity and the uncertainty principle is important because in his early presentations Heisenberg tended to interpret the uncertainty principle in a way which supported the idea that the task of science is merely to develop a mathematical formalism to predict observed phenomena. Such an "anti-realistic" view of science resigns itself from any attempt to understand the nature of the physical reality which produces such phenomena. Under such an interpretation of quantum theory, the statements of the theory do not make any claims about the nature of atomic systems themselves. Instead, the theory is justified solely because the task of theory is merely to predict successfully the phenomena which will be observed in specified experimental conditions.

As is well known, Einstein, whose opinion naturally carried great weight at the time, almost immediately expressed his distaste for both this indeterminism and what he took to be Heisenberg's anti-realist method of justifying his discovery. Since Bohr was first presenting his viewpoint of complementarity at precisely the same time as the appearance of the uncertainty principle, and since Heisenberg was closely associated with Bohr, and since Bohr saw the uncertainty principle as confirming his conviction that quantum theory was "complete", it seemed obvious to assume that Bohr's view coincided with what was commonly taken to be Heisenberg's anti-realism. Thus it appeared to Einstein that Bohr was defending the completeness of the quantum theoretical description by abandoning the description of physical reality as the goal of science.

As we shall see in the final chapters, Bohr's attitude towards science clearly rejected such an outlook.[21] While Bohr definitely argues against the form of realism defended by Einstein, he must defend a different form of realism if complementarity is to be a coherent framework for the description of nature. Thus complementarity offers a realist interpretation of scientific theory in the sense that it provides knowledge about what it takes to be independently existing real atomic systems and their constituents. Therefore Einstein definitely misunderstood the view Bohr was defending.

Nevertheless, it was Bohr who took upon himself the task of replying to Einstein's proposed thought-experiments. We will examine in Chapter Five just how this famous series of exchanges between these two scientific giants shaped the development of complementarity. At present my point is that these historical circumstances, coupled with the fact that at the time anti-realistic interpretations of science were very influential in philosophy, made it natural enough to assume that complementarity had abandoned the attempt to describe the physical reality which is the grounds of the phenomena of quantum theory. Certainly Einstein assumed as much and many of Bohr's opponents followed suit in this respect. Moreover, defenders of various anti-realistic philosophies of science were also glad to claim Bohr as an ally in their cause. There is no doubt that these facts provide the single most important reason for the prevalent philosophical confusion about complementarity and the lack of appreciation of Bohr's epistemological lesson.

To be sure, Bohr occasionally spoke in ways to abet this confusion—for he was naive in the language and arguments of the philosophers—and Heisenberg did so frequently. But a strict following of the historical record as well as a sympathetic understanding of Bohr's orientation as a natural philosopher should make it beyond doubt that neither did an anti-realistic outlook serve as the inspiration for complementarity nor was complementarity designed to give such an interpretation to quantum physics. As we have already seen, it was pursuing the implications that the classical framework must be made more "general" to incorporate the empirical discovery expressed in the quantum postulate that led Bohr to complementarity, not any prior philosophical commitment to any particular philosophy of science.[22]

When it is not recognized that complementarity is a framework designed to replace the classical framework, the other great quantum paradox, the "dualism of particle and wave descriptions", also suggests misreading Bohr as a foe of realism. It was reflection on the paradox that atomic processes could be represented as the effect of systems which are described sometimes as particle-like and sometimes as wave-like that led Bohr to recognize that there was a fundamental difficulty in the classical realist understanding of particle and wave concepts. Thus he concluded that the description of atomic phenomena through using these concepts could not be understood as a picture by which we can visualize what really happens in the interaction between atomic systems and the observing apparatus. Consequently he argued that we must renounce the classical expectation to be able to visualize an atomic system as it exists apart from those phenomena in which the physicist speaks of making an "observation" or "measurement" of some property of atomic systems. Thus Bohr was indeed a foe of the realistic understanding of particle and wave descriptions of phenomena as viewed from within the classical framework. He was, in other words, against the realism that Einstein seemingly wanted to defend, what might be called "classical realism". However, to conclude from this fact that he therefore embraced an anti-realist understanding of science would require assuming that there is no realistic interpretation of science other than that which operates from the viewpoint of the classical framework. Yet in offering a new framework complementarity is designed to provide for precisely such an alternative possibility. The anti-realist interpretation misses the central point that complementarity presents a revision of the classical framework which alters the options between anti-realism and a classical realist interpretation of physics. For this reason, the anti-realist misinterpretation makes complementarity appear to arrive dogmatically at *ad hoc* conclusions designed simply to save the existing incomplete quantum theory.

Bohr's acceptance of wave–particle dualism as unavoidable led early audiences to conclude that he was simply using the word "complementary" to refer to his conclusion that quantum mechanical systems must be described through two mutually exclusive, but complementary "pictures". Certainly Bohr did hold that both forms of representation must be used, but if this statement is taken as Bohr's alleged "principle of complementarity", then the substance of complementarity as a new framework is essentially lost from view. Nevertheless, on the basis of an anti-realist reading of theories, it was easy enough to construe Bohr's viewpoint as holding that there was no paradox involved in wave–particle dualism simply because both concepts are merely heuristic devices for predicting the observable outcome of particular experiments. A paradox arises only if we give these concepts a realistic interpretation and regard atomic systems as really having the contradictory properties of particles and waves. Bohr certainly meant to argue against the classical assumption that the same conceptual terms defined for describing observed phenomena can also be used for describing properties imagined to characterize an independent physical reality. But this argument does not imply that it is meaningless to posit the existence of an underlying physical reality to which the term "atomic system" refers. Nor does it establish the anti-realist conclusion that such theoretical notions are no more than conceptual instruments for predicting successfully the outcome of various experiments.

As we shall see in Chapter Four, complementarity does indeed teach that a new framework must involve a complementary combination of two distinct modes of description. But if these complementary modes of description are then simply equated to particle and wave pictures, it becomes difficult to see why we *must* use such dualistic concepts to describe atomic phenomena. One is naturally led to ask why it is not possible that an alternative way of representing atomic systems might be constructed such that all phenomena could be satisfactorily described by representing atomic systems through a single "picture" without requiring the paradoxical combination of two complementary pictures. If complementarity is interpreted in an anti-realist spirit, its conclusion that such a third way is not possible does seem to imply a rather dogmatic *ad hoc* limit on the human ability to create new descriptive concepts. Thus if we imagine that complementarity is only the principle that both wave and particle "pictures" must be used to describe atomic phenomena, then all of the labels of "arbitrary", "*ad hoc*" and "dogmatic" with which critics have branded Bohr's interpretation of quantum theory become defensible accusations.[23]

The reason for the common misreading of Bohr as an anti-realist lies not only in his attack against classical realism but also in his lack of any criticism of such an interpretation of quantum physics. But Bohr never attacked anti-realism not because he embraced this view but simply because he considered it so foreign to the basic presuppositions of natural philosophy. The nature of the philosophical debate between realism and anti-realism lay essentially outside of Bohr's intellectual horizons, whereas the whole debate to which he applied his energies was between classical realism and a possible new framework which would allow a realistic understanding of quantum physics by radically revising classical ideals of description. Realism was not the issue as Bohr saw it; the term itself was hardly part of his working vocabulary. He took it as empirically demonstrated that atomic systems were real objects which it was the goal of an acceptable atomic theory to describe. At least as Bohr understood it, the debate was joined over the nature of the framework within which the description of such objects is to be understood.

Philosophical misconceptions about complementarity often arise from Bohr's occasional unfortunate choice of expression. For example, in his early essays he occasionally spoke of observation as "disturbing" the state of a physical system. This way of speaking suggests that the limitations expressed by the uncertainty principle simply result from the fact that the physical act of observation "disturbs" the state of the observed system so that we cannot know it exactly. But this "disturbance interpretation" uncritically accepts the classical presupposition that the parameters used to define the classical state of the system refer to properties possessed by an independent physical reality. As we shall see, Bohr's revision of the classical framework requires that this very presupposition must be abandoned.

Another common misunderstanding of complementarity is that Bohr defends a subjectivistic interpretation of quantum theory. It is true that viewed from the criterion of objectivity in the classical framework Bohr's renunciation of the attempt to visualize an independently existing physical object in the atomic domain seems tantamount to abandoning the objectivity of scientific description. However, we saw in the previous section that as an alternative framework complementarity offers a new ideal of objectivity. As we shall see in Chapter Seven, it is a cardinal point for the framework of complementarity that the observations supporting scientific theory are *physical interactions* which the theory is intended to describe. Thus Bohr strongly emphasizes that the description of phenomena must take into account the effect of the "observer" on the observed object in the observational interaction. In the classical framework this effect could be "controlled", for it could be represented as arbitrarily small, but this is not possible in the quantum theoretical representation. Thus it is quite correct to say that Bohr insists on the incursion of the "observer" *as a physical system* into the description of phenomena. However, because he sometimes refers to this observer as the "subject", readers used to the philosophical meaning attached to this term have misunderstood Bohr as holding that quantum observations involve a mental–physical interaction. Since Bohr's epistemological lesson concerns how we can use concepts in the description of nature, it is accurate to say that quantum theory tells us what we can *know* about atomic systems. But this fact certainly does not entail that the quantum mechanical description of "observation" is a description of the "epistemic event" of coming to *know* that a certain interaction had a specific effect on the physical measuring system. Certainly such a claim was the furthest possible from Bohr's intentions, hence he had a hard time seeing how seriously he was being misread. When he finally understood this, he began to speak strongly about the fact that complementarity provided an *objective* means for describing observation, but he never overcame that early characterization of complementarity as subjectivistic.

The immense influence of Bohr's complementarity viewpoint, as evidenced not least in the historic debate with Einstein, naturally suggests asking why complementarity has been the subject of such frequent misinterpretations. Several distinct reasons probably contributed to the most important misunderstandings. First, as has been suggested in the above comments on the "disturbance" and the "subjectivist" misinterpretations, Bohr often misused key terms in his arguments. He was not well trained in the vocabulary or the problems of traditional philosophy and thus did not always understand how his assertions would sound to readers who did not share his outlook. Furthermore, partially due to the uniqueness of his position, he often adopted a less than self-evident idiosyncratic expression to refer to a crucial point. Although from youth he was indeed keenly aware of what would normally be called "philosophical problems", as we shall see in the next chapter, the formative influences which shaped his outlook cannot be said to have included any strong allegiance to a particular philosopher or school of thought.

Second, the nature of Bohr's writings did not lend itself to the presentation of his framework in any systematic detailed fashion. All of his philosophical work involved short essays, generally intended for public delivery to large audiences with diverse backgrounds. His three philosophical books are simply collections of these articles. Since each one was conceived as an independent piece, each one starts "from the beginning" over again. Such a means of presentation hardly allowed the sort of systematic treatment which a book-length exposition of complementarity would have provided. Bohr many times expressed the desire to write such a book, but the circumstances of his life never permitted it. One might also suppose that his method of work and his temperament were not of a sort that would have been ideal for such a task.

Third, there were historical factors entirely out of Bohr's control that created a climate conducive to misunderstanding complementarity. Not least of these was the fact that during the period of Bohr's philosophical work, most of those competent to write on the philosophy of physics were committed to some form of "positivist" philosophy of science. Many of Bohr's statements, particularly those designed to reply to Einstein's defense of classical realism, would sound perfectly in context with the anti-realist understanding of science that was often defended by positivists. In his mature years Einstein was clearly opposed to the positivists and Bohr was opposed to Einstein. Thus it was not uncommon to identify Bohr with this epistemology, a circumstance which was hardly conducive to the best understanding of complementarity.

The dominance of positivism in philosophy of science also suggests a fourth source of misunderstandings. Bohr certainly believed the atomic system was an independently existing physical reality; he regarded atomic physics as the attempt to gain empirical knowledge about these entities. His point was that we cannot describe these systems like classical physics described planets and billiard balls. In this spirit he repeatedly cautioned against using the classical mechanical concepts to try to describe independently existing atomic systems. Moreover he was extremely reluctant to comment about the nature of the atomic domain other than how we must describe the observational interactions through which it presents its phenomenal manifestations. Thus in spite of the fact that his move to a new framework implied revision in ontology, Bohr refused to venture into ontological discussions. The reason for this is probably to be found in the fact throughout the period of his formal education metaphysics was dominated by the German and British schools of idealism, a philosophy very foreign to Bohr's physical interests and against which positivism was a reaction. When Bohr began his philosophical writing in the 1920's and 30's, the ascendancy of positivism tended to make reference to the concept of physical reality taboo for scientifically minded thinkers. Because of Bohr's *partial* agreement with the positivists in opposing classical realism, it is not unlikely that their influence was a factor in his reluctance to explore the ontological consequences of complementarity.

One might point out that in spite of the de-emphasis on ontology in philosophy of science, certainly ontological issues were very much in the air at the time, for the whole quantum "mystery" was precipitated by wondering about what the nature of physical reality is like in the atomic domain. Nevertheless, Bohr's failure to recognize the ontological dimension of his thought was not a trivial omission. As we shall see in Chapter Five, the momentous debate between Bohr and Einstein was essentially a clash of viewpoints on an ontological level, but neither participant in the debate seemed explicitly to realize this fact. Certainly the disagreement between these two great natural philosophers did have an important role in clarifying the application of complementarity to a host of physical situations. But the two men were operating in different frameworks, thus placing different demands on what is required of an adequate description. Therefore, their chosen method of applying quantum theory to various thought experiments was unable to resolve their dispute.[25] The loss to philosophy has been the consequent obscuring of the philosophical significance of complementarity, particularly with respect to the concept of physical reality. As a result, philosophers have failed to recognize that when physical reality is understood from the viewpoint of complementarity, it is possible to retain a realistic interpretation of quantum theory and yet avoid the paradoxes that arise when a realistic interpretation is attempted from the classical viewpoint. Showing how this can and must be done is the concern of the final chapter, Chapter Eight.

3. The Philosophical Significance of Complementarity

The persistent confusion and misunderstandings about complementarity outlined in the previous section provide sufficient reason for a careful study of Bohr's viewpoint, but at least three additional reasons should be recognized. First, there is the purely historical point that in fact complementarity has had a tremendous influence on the development of twentieth century physics. As Professor Jammer points out, "In spite of the opposition to Bohr's views by some leading physicists like Einstein and Schrödinger the vast majority of physicists accepted the complementarity interpretation in general without reservations, at least during the first two decades after its inception."[26] Nevertheless, in spite of its dominance during this period of the awesome growth of atomic physics, the "textbook" presentations of complementarity which introduced most physics students to Bohr's views hardly could be considered to do the subject justice. As Jammer continues: "Usually at the end of a discussion of the Heisenberg relations a statement was added to the effect that the logical relationship between the concepts of position and momentum is called complementarity and in a footnote reference was made to some of Bohr's writings. In any case, all texts written between 1930 and 1950 ... espoused the complementarity principle even if they did not name it."[27] Even if it was only lip-service that was given to complementarity, surely an interpretation so widely professed deserves careful study.

A second point arises from the discussion of the interpretation of quantum physics which remains a burning issue in the foundations of physics. In dealing with a mathematical formalism which is in many ways more sophisticated than the quantum mechanics on which Bohr based his arguments, in the last decades physicists and philosophers have mounted numerous challenges to Bohr's so-called "orthodox interpretation". It is important to realize that none of these challenges have won the day or provided anything remotely like an empirical refutation of complementarity. Indeed the repeated failure to "disprove" complementarity alone suggests that any doctrine so durable is worthy of careful study. In any event, if debate between defenders and opponents of the so-called orthodox interpretation is to be carried on intelligently, the least one could demand would be an informed understanding of Bohr's position.

The third reason for an attempt to dispel the persistent misunderstandings of complementarity concerns its philosophical significance. Natural science may interest the philosopher either for its method of gaining knowledge or for its achievements in describing the natural world. But the two are not independent: if scientific theory is interpreted *realistically*, *i.e.*, if its claims are understood as aspiring to a true description of an independently existing order of nature, then its achievements tell us both something about the nature of that reality and about how we can come to have knowledge of it. As an achievement in describing at least one aspect of reality, the rise of modern atomic physics in which Bohr's outlook was so influential provides evidence for fundamental changes needed in the philosophical account of scientific knowledge.

Understanding the relationship between complementarity as a conceptual framework and quantum theory requires recognizing (as has been recognized in much recent philosophy of science) that the acceptance of a theoretical account of specific phenomena is at least partially determined by its conformity with the scientists' expectations for any description of that sort of phenomena.[28] The descriptive ideals which operate as criteria for the acceptance or rejection of a theory are implicitly stipulated by the framework within which any theoretical description is offered for acceptance. The resistance to accepting the theoretical descriptions of quantum mechanics is not caused by any lack of empirical confirmation, but must be attributed to its failure to conform to the descriptive ideals laid down by the classical framework. This lack of conformity produces the apparently paradoxical quality of quantum mechanical descriptions. Because Bohr very strongly believed that through its spectacular empirical confirmation, quantum theory had proved that it was here to stay, he argued that accepting quantum theory implied abandoning classical descriptive ideals. One could, of course, simply reject quantum theory and thus retain the classical criteria for an acceptable description. This was, in a sense, the route taken by Einstein. But since Bohr believed that an alternative framework was available in complementarity and that this new framework removed the paradoxical character of quantum theory, any retention of the older framework seemed to him a dogmatic obstacle to the further understanding of nature.

As a new conceptual framework complementarity stipulates essentially new ideals for describing atomic phenomena. In doing so, it changes the presuppositions for the application of theoretical concepts to the description of nature. Bohr's intention in proposing this new framework is to show that it makes possible a unified, consistent, harmonious description of all the relevant phenomena. His arguments to show how this framework permits describing phenomena teach an epistemological lesson about the use of concepts to describe phenomena in general.

Although failing to make possible a harmonious and unified description of nature will precipitate the downfall of any framework for physical science, history seems to teach that such a passing will take place only when a new framework is present to make possible a more adequate view. Bohr believed such a framework was available and that he had found it in complementarity. Einstein did not. Thus to Bohr, Einstein's clinging to the classical framework seemed pure dogmatism; whereas to Einstein, Bohr's insistence that quantum theory provided a complete description of atomic phenomena seemed equivalent to renouncing a primary goal of scientific description.

This chapter began by seeking the genus of complementarity, what sort of a thing is it? I answered that it is a conceptual framework for the description of nature in physical science, specifically atomic physics. But if we consider complementarity to be the whole of Bohr's philosophical contribution, then complementarity includes not only the framework itself, but also the arguments for generalizing the classical framework to produce a new one, the epistemological lesson that adopting this framework teaches about the presuppositions for the use of concepts, and the ontological repercussions generated by viewing physical reality in the atomic domain from this framework. It seems appropriate to call this conjunction of ideas a "philosophy" and to speak of Bohr's philosophy of complementarity. Bohr's thought habitually moved outward to trace the logical consequences which changes in the presuppositions for using a particular concept would have on how we regard the description of nature. Therefore, due to the unity of nature, any restructuring in the presuppositions for the use of concepts at one level may well teach a lesson concerning the concepts used to describe observation in other fields of science as well. For this reason also, "philosophy" is appropriate to designate the genus of Bohr's thought.

To be sure, the ideas Bohr put forward under the name of complementarity lack the structure which I shall try to impart to them in the following chapters. His terms are often undefined and his arguments frustratingly obscure. The detailed application of his ideas to describing phenomena outside of atomic physics is sadly lacking. And there are serious omissions from both an epistemological and ontological point of view. However, other intellectual creations have suffered as grievously from such flaws and still earned the label "philosophy". Only if Bohr's whole philosophy is understood and his new viewpoint adopted will his interpretation of quantum theory have any logical roots sunk in the fundamental character of human knowledge and the scientific description of nature. This study is dedicated to making such a case.

CHAPTER TWO

BOHR'S PHILOSOPHICAL ORIENTATION

In what was to be his last interview, the day before his death, Bohr was questioned by Thomas Kuhn about the nature of his interest in fundamental philosophical problems. His answer was direct: "It was in some ways my life, you see."[1] Indeed Bohr was truly what would have been called in an earlier day a "natural philosopher". More than anyone else, he helped to shape and transform atomic physics from an unknown frontier to a well developed and explored terrain. Consequently as a scientist he was constantly engaged in fundamental problems concerning questions once considered the province of philosophers. To attempt to understand complementarity from the standpoint of the *completed* quantum theory without considering how Bohr was peculiarly prepared to formulate this framework would inevitably distort our story and render it mysterious. We will follow the course of Bohr's contributions to quantum theory itself throughout the next chapter when we will directly examine his journey to complementarity. In this chapter we will consider only his preparation for that journey in terms of his early education and methods of work. However, before turning to these topics. a biographical sketch of the main events of his life will provide a useful framework within which to consider the historical development of complementarity.

1. Biographical Sketch of Bohr's Intellectual Life

Niels Bohr was born on October 7, 1885 in Copenhagen, the city whose name he was to link forever with the development of atomic physics. His father, Christian Bohr, was a well-known professor of physiology at the University of Copenhagen; his grandfather and maternal aunt were also important figures in Danish educational institutions. Bohr's immersion in the pleasant, secure life of the upper middle-class intelligentsia allowed the best possible circumstances for his talents to be recognized and cultivated to the fullest.

Bohr's Danish heritage may seem inconsequential, but everyone familiar with him knew full well that his nationality was no trivial matter. Except for the dark days of World War II the small and politically unimportant Denmark provided a secure environment conducive to scientific and philosophical reflection. Also the political neutrality of Denmark throughout World War I and until 1940 in World War II enabled Bohr's Institute in Copenhagen to serve not only as a meeting ground for international scientific collaboration on an unprecedented scale, but also as a refuge instrumental in saving the lives of many of the century's greatest scientists.

Culturally Denmark stood midway between British and Germanic traditions, allowing in a real sense a fortunate synthesis of British experimental science with the more formal, theoretical approach of the German Universities. In many ways Bohr's philosophical temperament combined British influences stemming from the Lockean tradition of common sense empiricism with the typically German heritage of Kantian concerns with subjective and objective aspects of experience. With respect to his philosophical development, Bohr's Danish heritage allowed him a certain freedom from the overruling "schools" of idealism and materialism, closely identified with various national traditions. (Later in this chapter we will consider the possible influence of the one great Danish philosopher, Søren Kierkegaard.) Late in life when Bohr's fame was international, the intimacy of this small land brought him the status of virtually a folk hero, generally accorded only to those in more publicly conspicuous fields. Today sculpted busts of Bohr adorn The University and the City Hall of Copenhagen in company with two other Danish cultural heroes, Kierkegaard and Hans Christian Anderson.

Bohr had one elder sister, Jenny, and a brother, Harald, one and a half years his junior. The relationship between Niels and Harald was remarkable and had an important influence on Niels' method of working. From childhood the two brothers learned to express and sharpen their ideas in vigorous dialogue, thereby developing an incredibly interdependent dialectical rapport with each other's patterns of thought. The amazement of witnesses to this phenomenon, similar to Socratic dialogues, is recorded in this typical reminiscence by one of their schoolmates:

> Their way of thinking seemed to be co-ordinated; one improved on the other's, or his own expressions, or defended in a heated yet good-humored manner his choice of words. Ideas changed their tone and became polished; there was no defense of preconceived opinions, but the whole of the argument was spontaneous. This way of thinking *à deux* was so deeply ingrained in the brothers that nobody else could join it.[2]

This introduction to the art of argumentation conditioned Bohr to need to work out his ideas in companionship with another individual. Except on rare occasions he always found it necessary to engage in such a dialectical interchange in order to carry through his philosophical reflections as well as his revolutionary scientific work. The relationship with Harald, who later attained world-wide eminence as a mathematician and directed the mathematics institute next door to Niels' Institute for Theoretical Physics, provided Niels with a living resource of mathematical knowledge necessary for his work, as well as a vital bond from which he derived inspiration and encouragement.

Throughout childhood Harald was the more precocious of the brothers; though the younger, he finished his master's degree before Niels and traveled to Göttingen, then an international center for mathematics. In spite of Harald's achievements, the boys' father singled out Niels as "the special one of the family", noting his exceptionally deep method of thinking was characterized by a persistence in making clear the most fundamental principles underlying any attempt at an explanation.[3] In Chapter Three we will see how this persistence contributed to the development of quantum theory and complementarity.

In the spring of 1911 Niels finished and defended his doctoral dissertation on the electron theory of metals. Then it was customary for a recent doctoral graduate to do postdoctoral research outside of Denmark. In continuation of his dissertation work on the electron theory, Bohr quite predictably chose to go to Cambridge where the famous Cavendish Laboratory was directed by J.J. Thomson, discoverer of the electron. There is no evidence that Bohr was seriously unhappy at Cambridge, but because of changing interests Thomson failed to appreciate the talents of this young foreign student whose spoken English was often inadequate to the necessities of precise argument. Nevertheless, Bohr did familiarize himself thoroughly with Thomson's work on atomic models and came to understand their fundamental inadequacy. The year before Bohr's arrival in England, the New Zealander, Ernest Rutherford, had made the startling discovery of the atomic nucleus, thereby precipitating the fame of his department at Manchester. Bohr met Rutherford sometime in the fall of 1911; undoubtedly both were impressed with each other, for by April of 1912 Bohr left Cambridge for Manchester.

The spring and summer of 1912 was an astonishing time for the twenty-six year old Dane. During this period Bohr came to recognize that Rutherford's discovery of the atomic nucleus necessarily would be the cornerstone for building any model of the atom; yet he also realized that in classical physics no such atom could ever be represented as a mechanically stable system. At the same time he became increasingly clear about the fact that the quantum of action must play a role in the development of any theory of the atom. However, his first stay in Manchester was brief, for it ended with Bohr's return to Denmark to marry Margrethe Nørlund on August 1, 1912. That fall and the spring and summer of 1913 Bohr produced a series of three papers in which he outlined his revolutionary new theory of the atom and applied the postulates of his model to the formation of atomic spectra. But he made no claim to provide a "final" description of atomic systems. The break with classical physics which Bohr advanced in his theory was so great that to some his papers appeared to be virtually numerological trickery, but the power of the theory to derive empirically confirmed relationships demanded awesome respect. Thus while no flood of applause burst forth with the first appearance of his papers, recognition built steadily. We will consider the contents of these papers in the next chapter; at present our only concern is with fixing the biographical chronology and familiarizing ourselves with Bohr's approach to scientific and philosophical problems.

Bohr's fundamental atomic theory appeared in three parts in *The Philosophical Magazine* through the summer and fall of 1913.[4] In 1913–14 he gave an informal course of lectures at The University of Copenhagen, while the decision to grant him a professorship was pending. During this year he made several trips to England, and in September attended a vital discussion of his theory in Birmingham at the annual meeting of the British Association for the Advancement of Science. At this meeting the announcement of new experimental evidence improved the reception of the theory among the originally skeptical British audience. Although the misgivings of the British scientists were put to rest by the improving relation between the theory and experiments, the German mathematicians of Göttingen gave his ideas a cool reception because they distrusted Bohr's application of the mathematics of classical physics to a model which defied the very basis of the classical view. In July a trip to Germany helped him to win support from that quarter, including the very important convert of Max Born, later to provide a crucial link in the development of the theory.

In the spring of 1914, Rutherford offered Bohr a readership at Manchester for the year of 1914–15, later extended to 1916. In May of 1916 he finally received appointment as professor in theoretical physics in Copenhagen. (Although at this time he was invited to lecture at the University of California at Berkeley, Bohr could not make good his wish to visit the United States until 1923.) In the autumn of 1916 he was joined by his first assistant, the Dutch physicist H.W. Kramers, who was to stay with Bohr in Copenhagen until 1926. In the spring of 1918 Oskar Klein became his second student assistant. In 1917 he began negotiations to build a laboratory as the foundation for a new "Institute for Theoretical Physics", but it was four years before the Institute was to open its doors. Through these doors was to pass one of the most astonishing series of brilliant scientists, as students, colleagues, and guests, that the history of science has ever witnessed.

Throughout the ten years following his original 1913 papers Bohr continued with the help of his assistants to extend his theory of atomic structure by applying it to the entire periodic table of elements. The last part of this great work was added in 1922. That year marked a fundamental change in the nature of Bohr's activities, for in June he delivered a series of lectures in Göttingen, facetiously called the *Bohrfestspiel* where he first made the acquaintance of two very young physicists, Wolfgang Pauli and Werner Heisenberg, who were to be in and out of Copenhagen for the next five years and who were to be important among those who revolutionized Bohr's own revolution. As Oskar Klein recalled, by now Bohr assumed a new role in the advance of atomic physics:

> Bohr who had earlier met with considerable criticism and lack of understanding, had at this time become one to whom all listened with reverence, so that the discussions about the lectures were rather concerned with whether Bohr had meant this or that, than the matter itself....
> ... The time now approached when the main progress was to come from scientists of a younger generation, but with Bohr still as a giver of ideas and increasingly as the philosophical integrator of the knowledge obtained.[5]

These marvelously productive years were given an elegant conclusion by Bohr's reception of the Nobel Prize in physics in December of 1922.

Throughout the decade of the twenties Bohr moved increasingly into the role of Director of the ever expanding Institute which became more and more the ultimate clearinghouse for all progress in atomic physics. However, this change of roles did not mean that he believed that the fundamental revolution in physics initiated by his atomic model had now settled down, leaving only the mathematical detail to be elaborated by the younger men. Indeed, this decade is the primary one in the story of complementarity, so a somewhat more detailed listing of its important events is in order.

In fact, as the exciting breakthroughs of quantum mechanics appeared, Bohr seemed to greet each with a certain ambivalence, applauding the dramatic progress being made, but grieving over the inconsistencies between classical and quantum theories. In collaboration with Kramers and John Slater, Bohr published in 1924 what was to be the last attempt to describe the atomic system along the quasi-classical lines he had employed earlier. Although the suggestion of this paper that energy is not strictly conserved in individual atomic interactions was quickly refuted by experiment, the revolutionary character of this proposal indicates how desperate Bohr considered the situation at this time. The same year Kramers succeeded in formulating a mathematical theory for the dispersion of light by atoms. Working from Kramers base, in the summer of 1925 Heisenberg developed a purely abstract mathematical representation of quantum mechanical systems. Throughout 1925–26 Heisenberg refined and extended his theory, with the help of Max Born and Pascual Jordan, to produce what is now called "matrix mechanics". That spring Erwin Schrödinger working quite independently put forward a "wave mechanics" representation of quantum systems, later discovered to be mathematically equivalent to Heisenberg's matrix mechanics. These two different approaches convinced Bohr that mathematically their theory was on the right track, but at the same time they intensified his deep concern with the physical interpretation of the mathematical formalism. On the one hand Bohr joyfully welcomed the new theories as furthering the progress of atomic physics in its ability to predict confirmed results for an increasing range of experimental phenomena. But on the other hand he was made ever more aware of that tension which he had felt since 1912 created by the fact that quantum theory had thrown out the classical framework but still had need of a framework of its own to stipulate what could be expected of a quantum mechanical description of atomic systems. Thus by the summer of 1926 Bohr was still very much in the vanguard of the revolution which his theory had begun. Indeed, more than any other physicist, he intuitively felt the the inadequacy which the lack of a physical interpretation of the theory gave his approach of combining classical and quantum ideas. Heisenberg recalls in quite dramatic terms the intensity of Bohr's agony at this time:

> Bohr was more worried than anybody about the inconsistencies of quantum theory. So he tried really to understand what is behind these difficulties. ... Bohr really suffered from it, and Bohr couldn't talk of anything else. ... He in some ways directly suffered from this impossibility to penetrate into this very *unanschaulich*, unreasonable behavior of nature. ... But that was Bohr's whole philosophical attitude—he was a man who really always wanted to get the last degree of clarity. He would never stop before the end. ... Bohr would follow the thing to the very end, just to the point where he was just at the wall. ... He

> did see that the whole theory was on the one hand extremely successful, and on the other hand was fundamentally wrong. And that was a contradiction which was very difficult to bear, especially for a man who had formulated the theory. So he was in a continuous inner discussion about the problem. He always worried, "what has happened?"[6]

It was only with great intellectual torment that Bohr began to grope his way towards complementarity.

In 1926–27 Heisenberg returned to the University of Copenhagen to discuss the pressing issues with Bohr. That fall Erwin Schrödinger's visit to the Institute precipitated a new level of intensity in the discussions on the physical basis of the new theories. While Bohr and Heisenberg had always approached atomic physics from the direction of treating atomic systems as pseudo-mechanical configurations of particles, Schrödinger treated the atomic system as a wave phenomenon. These discussions convinced Bohr to take seriously the wave–particle dualism of light phenomena in the interpretation of atomic systems. Working in Copenhagen in February of 1927, Heisenberg isolated and formulated the now famous uncertainty principle. At the same time, while on a skiing holiday in Norway, Bohr began to put together the basis of the new conceptual framework of complementarity. Though Bohr first greeted the uncertainty relations with some skepticism, by the late spring of 1927, intense discussions between Bohr and Heisenberg, mediated by Klein, managed to convince all participants that the two approaches were harmonious and could serve as a basis for the arguments of complementarity.

Complementarity was first revealed to a public audience in September of 1927 at a congress in Como, Italy. Although Bohr had worked all summer on the manuscript, the delivered paper was far from finished form. Most of the audience was unimpressed, finding Bohr's argumentation far too "philosophical" and including nothing new in physics. But Pauli recognized the significance of the new ideas and worked with Bohr in Como after the conference to produce a more refined manuscript. Even that, however, was far from the finished product, for throughout the fall of 1927 and the following spring the manuscript went through repeated rewritings. Finally it was ready for print by Easter of 1928.

Meanwhile in October of 1927 Bohr had an opportunity to present complementarity a second time at the Solvay Congress in Brussels. Here were gathered the "greats" of European physics, including Einstein who had not been at Como. Einstein's negative reaction was a grave disappointment to Bohr, but it produced the series of discussions to which I have referred in the previous chapter. These discussions culminated in 1935 in the exchange of articles between Einstein and Bohr on the so-called "Einstein–Podolsky–Rosen paradox", and for good or for ill shaped the way Bohr was to refine and present complementarity. In Chapter Five we shall examine the effect of this interchange.

Throughout the thirties Bohr divided his efforts between extending and refining complementarity, and carrying out new physical investigations in electrodynamics and nuclear models. His first collection of articles on complementarity, including the Como paper of 1927, was published in German in 1931 and in English in 1934. This work, *Atomic Theory and the Description of Nature*, will form the basis of our examination of complementarity in Chapter Four.[7] Except for a paper co-authored with his colleague, Leon Rosenfeld in 1933 applying complementarity to the measurement of fields, Bohr's research at this time does not bear on our concern with complementarity; so we shall not discuss it here.

Prior to the occupation of Denmark, Bohr had worked incessantly to secure the escape of many a European scientist from the spreading Nazi horror. In 1943 the Nazis occupied Bohr's Institute and shortly afterwards, under threat to his life, Bohr was forced to flee Denmark for neutral Sweden; shortly afterwards he was flown to England. In November of 1943 Bohr traveled to America as part of the British team working on the development of the atomic bomb. He had made vital contributions to the theory of nuclear fission in 1939 in collaboration with J.A. Wheeler at Princeton. But Bohr was convinced that a technologically sustained fission was impossible. Once back in America, he spent much of his time until the end of the war at Los Alamos where he served as a valuable resource person, though the direction of the work was left to others.

Throughout the forties and fifties until his death in 1962, Bohr increasingly turned his attention to the problems brought on by World War II and the subsequent social and political problems resulting from atomic fission. Once the war made inevitable what Bohr had previously thought impossible, he set about trying to prepare political and military leaders for the change in international relations which the advent of the atomic bomb would bring. A perpetual optimist, Bohr saw in the bomb the ultimate argument for disarmament and an open world. Only five days after the bombing of Hiroshima, he published these ideas in an article in *The Times* of London entitled "Science and Civilization".[8] Political leaders, primarily Churchill, were not impressed by Bohr's arguments. He attempted to take his case to a more public forum with an open letter to the United Nations containing a plea for an open world.[9] Presented in 1950, this entreaty appeared at a hostile moment, but Bohr never tired of putting his bounteous energies to the task of securing international scientific co-operation.

After the war Bohr became increasingly the world spokesman for nuclear and atomic physics; during this same time the nature of his profession altered dramatically. Theoretical physics became the interest of governments and big business and no longer the research of a few university professors in close communication. The costs of fundamental research assumed proportions made possible only by government support. Bohr's Institute expanded rapidly to keep up, though it never became a center for large-scale experimental work. As its Director, Bohr was increasingly drawn into administrative responsibilities as well as into the celebrity spotlight of the scientific world.

Throughout his last twenty-five years Bohr pursued the theme of complementarity as an epistemological lesson into a variety of fields in the description of natural phenomena. His finished public lectures on "atomic physics and human knowledge" are collected in two thin volumes.[10] In addition there are about seven separately published articles appearing in isolated journals. He gave two large-scale series of lectures on complementarity, the Gifford Lectures of 1949, delivered in Edinburgh, and the Karl Taylor Compton Lectures of 1957, delivered at M.I.T. Both series were intended to be published, but in spite of his professed desire to write a book-length exposition of his philosophy, he could never bring himself to rewriting and polishing the lecture transcripts.

In 1960 the American Physical Society and the American Philosophical Society jointly supported a project recording the history of quantum physics, primarily through interviews with the surviving physicists. In the spring of 1961 Professor Thomas Kuhn was appointed director of the project and received an invitation from Bohr to make the Institute in Copenhagen the European headquarters of the project. An extended series of interviews with Bohr were projected, but on November 18, 1962, the day after the fifth interview, Bohr died during an afternoon nap. In that last interview his emphatic concern with fundamental conceptual problems was as clearly in evidence as it had been at the very beginning of his work on atomic theory fifty years earlier.

2. Bohr's Method of Work

Bohr's manner of working is so distinctive that it has received much comment in the memoirs of those who worked with him. Typically, at the beginning of any project, Bohr started with the intention "to write a little paper on it". This would be accomplished by his dictating, often in a mixture of English, Danish, and German, to a student or co-worker (or, in earlier days to his mother or his wife) sentence by sentence the text of the paper to be. Oskar Klein, who was the scribe/respondent for the Como paper in which complementarity was first presented, recorded his initiation into this method:

> With some writing paper and a pencil in front of me I was placed at a table around which Bohr wandered, alternatingly dictating in English and explaining in Danish, while I tried to get the English text on paper. Sometimes there were long interruptions either for pondering what was to follow, or because Bohr had thought of something outside the theme which he had to tell me about.[11]

These dictations were not intended to produce directly a finished manuscript but to give concrete expression to the first move of a dialectical process of creation. Oddly enough the actual draft seemed irrelevant to Bohr, whose prodigious memory allowed him to recall the dictation almost verbatim without reference to the work his assistant had transcribed. This opening move would then result in innumerable rewritings, corrections, and additions. Leon Rosenfeld, Bohr's closest collaborator for the last half of his life, describes the process as follows:

> So first a draft is outlined on which criticism is at once brought to bear; innumerable retouchings and corrections soon cover the pages, fill the margins, insinuating themselves between the lines. But suddenly all this painfully constructed composition is abandoned: another way of looking at the matter has occurred to him and a fresh start is made along this new line. A second text is elaborated, embodying perhaps some remnant of the first, and is immediately subjected to the same criticism.[12]

This manner of composition naturally enabled Bohr to improve his statement, producing a well-formulated presentation, but unfortunately the final product buries all previous arguments, thereby hiding the dialectical process which led to its creation. For this reason positions which may have seemed perfectly clear to those who witnessed or participated in the writing of

Bohr's essays may seem opaque to those who have only its final manifestation before them. Because most alternative positions never make it to the final published essay, it may appear that Bohr has seized upon a single view and chosen to defend it tenaciously; however, this interpretation would be a serious misunderstanding. Though somewhat given to excessive statements about Bohr, Rosenfeld comments on the *openness* which is characteristic of Bohr's approach:

> When in full action Bohr's mind, with its restless probing, its adventurous explorations in all directions, reminds one of the ant's feelers, obstinately palpating on all its faces an obstacle suddenly encountered. Bohr likes to claim for the investigator of nature the unrestricted right to let "his thoughts go to and fro". These successive attempts slowly cover the problem with an ever tightening net; none of its aspects can escape its unceasing search.[13]

While this description may well be true, Rosenfeld misleads us in claiming that Bohr's style "endeavors to express a whole variety of aspects and points of view successively recognized, with all the maze of mutual connexions and interactions between them".[14] So it may have appeared to those who knew all each manuscript involved. But those who have only the finished product to examine are more likely to wish for a clearer account of how certain positions were reached.

It is an interesting measure of the seriousness with which Bohr approached his writing that when he sent the first draft of his atomic theory paper to Rutherford in Manchester, the latter was impressed, but feared the paper was much too long and begged leave to cut it down. Bohr immediately traveled from Denmark to Manchester to defend every word of the paper. Rutherford finally capitulated to any reduction in length, explaining, "I could see that he had weighed up every word in it, and it impressed me how determinedly he held on to every sentence, every expression, every quotation: everything had a definite reason, and although I first thought that many sentences could be omitted, it was clear, when he explained to me how closely knit the whole was, that it was impossible to change anything."[15] This weighing of each sentence, indeed each word, is very typical of Bohr's method and is repeatedly confirmed by all who worked with him. Yet so unusual is Bohr's position that even to a very knowledgeable reader like Rutherford, this fact is hardly obvious on a first reading.

It is also typical of Bohr's method, already visible in the atomic theory papers of 1913, that the mathematical derivations are never as important as the logical analysis of the concepts through which the phenomena are described. This tendency was to serve Bohr well in formulating complementarity, but it was atypical for scientific papers and gives them a relative obscurity to those more used to dealing with physical problems in the language of mathematics. Heisenberg, who was much more at ease with mathematical argumentation, and for this reason often differed with Bohr, insightfully recalls his impressions of Bohr's approach at their first meeting:

> When he explained the individual assumptions of his theory, he chose words very carefully. ... And each one of his carefully formulated sentences revealed a long chain of underlying thoughts, of philosophical reflections hinted at, but never fully expressed. ...
>
> ... Behind every carefully chosen word one could discern a long chain of thought, which eventually faded somewhere in the background into a philosophical viewpoint which fascinated me. ...
>
> ...His insight into the structure of the theory was not a result of mathematical analysis of the basic assumptions, but rather of an intense occupation with the actual phenomena, such that it was possible for him to sense the relationships intuitively rather than derive them formally.
>
> Thus I understood: knowledge of nature was primarily obtained in this way, and only as the next step can one succeed in fixing one's knowledge in mathematical form and subjecting it to complete rational analysis. Bohr was primarily a philosopher, not a physicist, but he understood that natural philosophy in our day and age carries weight only if its every detail can be subjected to the inexorable test of experiment.[16]

Bohr's rejection of purely mathematical derivations in favor of conceptual analysis of the physical basis on which scientific description rests grew out of his concern with the relation between descriptive concepts and the role of observation in describing nature.

This method of approach involves linking one problem to another so that correlations between hitherto unrelated phenomena emerge in a new light. Characteristic of Bohr's great papers is an almost perverse refusal to isolate one problem as the center of discussion and an apparently obstinate insistence on piling up one problem on top of the next. Bohr's constant return to the same fundamental issues over and over again is in startling contrast to an approach which adopts a set of assumptions and proceeds to deduce their logical consequences in a straightforward manner. Indeed, Bohr's work was such that no axiomatic assumptions could be used as any starting point, for what was required was not extending existing theory into a new domain, but rather defining the areas where the existing theory was deficient. For such a task Bohr's approach of starting and restarting over and over again allowed him to feel out the limits of the framework of classical physics and thereby have a sense of the reality that lay beyond its grasp. Rosenfeld points out that "the real clue for understanding the underlying motives" of Bohr's method is to be found in "his predilection for the features of natural phenomena illustrative of general and fundamental laws, and that characteristic endeavor to achieve a rational synthesis between apparently contradictory elements".[17] Bohr's strong urge towards synthesizing various points of view in order to achieve a framework adequate to describe all phenomena does not easily lend itself to presentation as an argument involving a single set of logically consistent assumptions. Rosenfeld observes that, "logical analysis was not for him a mere verification of consistency (which he regarded as trivial), but a powerful constructive tool, orienting the groping mind in the right direction".[18] But once one direction had been explored in a preliminary manner, Bohr typically begins again from a new perspective, thereby appearing to shift starting positions. The unity of Bohr's account cannot be found in its starting points but is seen in its destination, towards expressing an almost "intuitive feeling" for the phenomena in question. Rosenfeld describes this aspect of Bohr's outlook as follows:

> This imperious need for a solid grasp on reality provides ... an adequate compensation for certain northern mistiness, a love of half-tints, resulting from the equally strong urge that no shade of this infinitely complex reality be lost in the process of analysis. ... He keenly feels the insufficiency of any one-sided analytical procedure: the harmony of things is made up of the interplay of apparently conflicting aspects.[19]

As we shall see in the following chapters, this type of approach seems very appropriate for complementarity which insists that a logical analysis of the conditions required for unambiguous use of descriptive concepts requires employing distinct but complementary modes of description.

3. *Early Philosophical Influences*

In Bohr's last interview he mentions that he became seriously interested in philosophical problems, and even considered writing a paper on epistemology, while a student at the University of Copenhagen (1903–1907). There is good reason to believe that these early philosophical ideas may have had a role in shaping complementarity, for when he unveiled his "new viewpoint" some twenty years later, friends who had known him as a student remarked that this was nothing but what he had been saying all along.

Unfortunately it is quite impossible to reconstruct precisely the lines of Bohr's early ideas in epistemology because we must deal with vague and incomplete records of his student interests. Therefore great care should be used in attempting to isolate the philosophical influences on the development of complementarity. We may well agree with Professor Max Jammer's remark, "It is certainly true that philosophical considerations in their effect upon the physicist's mind act more like an undercurrent beneath the surface than a well-defined guiding line. It is the nature of science to obliterate its philosophical preconceptions, but it is the duty of the historian and philosopher of science to recover them under the superstructure of the scientific edifice."[20] Yet it is well to caution that it serves little purpose to attempt such reconstruction beyond what the historical record can substantiate.[21]

Attempts to detect philosophical influences on Bohr should be preceded by noting two points. First there is the question of Bohr's philosophical education. Apparently his sole mentor in this area was the man who was the then dean of Danish philosophers, Harald Høffding. Bohr attended Høffding's course on the history of philosophy, a field in which Høffding was certainly an eminent scholar; however, this course was no more than a one year, required introductory survey for beginning students.[22] One must realize that the depth of understanding of historical positions, terminology, and philosophers attainable in such a course is certainly limited. Thus it is doubtful whether as a young man Bohr had any more than a "textbook" awareness of the general outlines of the systems of the great philosophers. Throughout his life he remained relatively indifferent to attempts to connect complementarity to trends or themes in the history of philosophy. In fact his whole attitude could be summarized by the remark in his last interview that during his student days he "felt that philosophers were very odd people who really were lost". Concerning the analysis of problems in traditional Western philosophies he commented that he "felt that these various questions were treated in an irrelevant manner".[23] Klein confirms this conclusion with his remark that Bohr "read I think some of Høffding's books, but I think he read very little more".[24]

Second, with respect to his scientific work, Bohr always scrupulously recorded his intellectual debts. It seems highly unlikely that he would take a radically different attitude towards any supposed philosophical debts to well known philosophers. Yet in his published works as well as his unpublished manuscripts and letters he never refers to a single philosopher as having helped him on the road to complementarity. In fact, quite the contrary, he generally stresses that complementarity represents something entirely new in philosophy. Agreeing with this assessment is Rosenfeld's comment (as reported by Professor Gerald Holton) that "after discussing his early philosophical meditations and his pioneering work of 1912–13, he told me in an unusual solemn tone of voice, 'you must not forget that I was quite alone in working out those ideas and had no help from anybody' ".[25] Thus we can best answer the question of direct philosophical influences of which Bohr was aware by saying that there were none.

Of course this conclusion does not preclude the possibility that indirect influences acted as intellectual catalysts. Thus we must consider whether any possible channels of influence may have shaped his philosophical development. In this respect the historical record seems to substantiate two possibilities: 1) the discussions of his father, Christian Bohr, with Høffding and others of that generation of professors at The University of Copenhagen, and 2) the discussions of the philosophical club of Høffding's students, the Ekliptika Circle. We shall consider these in turn.

Bohr's father received a medical degree, but never practiced as a physician. His interests focused on the problem of the proper description of physiological processes and the resulting controversy between mechanistic and teleological modes of description. Christian Bohr discussed these questions with his long-time friend, Høffding, in groups that met regularly at the Bohr home. The young Bohr brothers were permitted to audit these sessions. His father's position was conditioned by the then current anti-materialistic reaction against the mechanistic materialism of Haekel and his school. The elder Bohr rejected the attempt to formulate strictly mechanistic descriptions in biology and insisted on the need for vitalistic or teleological accounts of physiological processes. Niels tells us directly that this issue of mechanism versus teleology was important in his own intellectual development. In an essay written in 1957 he quotes approvingly the following extended passage from his father's work:

> As far as physiology can be characterized as a special branch of natural science, its specific task is to investigate the phenomena peculiar to the organism. ... It is thus in the very nature of this task to refer the word purpose to the maintenance of the organism and consider as purposive the regulation mechanisms which serve this purpose. Just in this sense we shall ... use the notion of purposiveness about organic functions. In order that the application of this concept in each single case should not be empty or even misleading it must, however, be demanded that it be always preceded by an investigation of the organic phenomenon under consideration sufficiently thorough to illuminate step by step the special way in which it contributed to the maintenance of the organism.[26]

The younger Bohr understood his father to be claiming that when the descriptive concepts are precisely defined, there need be no collision between purposive and mechanistic descriptions. Indeed he argues that inasmuch as purposiveness is experienced as a characteristic of phenomena to be explained in biology, such a concept cannot be purged from the descriptive vocabulary of physiology if biological descriptions are to be true to experience. This concern with the empirical situation in which concepts become applicable to the description of phenomena becomes a hallmark of complementarity.

We should also note that these discussions of mechanism versus vitalism were not posed in terms of the question of the possible existence of some non-physical entity, a soul, psyche, or *elan vital*. In this respect neither the father nor the son were inclined to reject the metaphysical commitments of the world-view of classical science. What was rejected was the sufficiency of a purely mechanistic *mode of description* of physiological processes. Christian Bohr points out that a mechanistic analysis of an organ within an organism is applicable to a description of the behavior such an organism exhibits only when we specify and understand the circumstances in which we experience the phenomenon to be described. In this respect, looking back on his father's work from the perspective of complementarity many years later, Bohr saw a foreshadowing of his own position.

This exposure to the mechanism versus vitalism controversy almost certainly conditioned the young Bohr's approach and led to his rejecting the usual line of argument of the philosophers. He did not see the problem in *ontological* terms, whether or not the living organism is the sort of entity capable of being described purely mechanistically. Rather he understood the problem as concerning the applicability of specific concepts to specific phenomena. In other words, from the beginning Bohr's awareness of the task of science was focused on the questions of the description of phenomena and the development of a conceptual framework adequate for such a description.

The influence of Høffding and the Ekliptika Circle is much harder to assess. As a young man Høffding had been fascinated by the philosophy of Søren Kierkegaard and did much to spread Kierkegaard's reputation outside of Denmark. Ultimately Høffding took his leave of Kierkegaard as a theologian, but what remained of Kierkegaard's influence on Høffding was the former's rejection of the Hegelian attempt to enclose all reality within a single embracing conceptual system. This outlook conditioned Høffding's reading of the history of philosophy, and it is reasonable to conclude that such an attitude was passed on to the young Bohr.

Other than this general anti-absolutist attitude towards speculative philosophy, it is difficult to impute to Høffding the position of any philosophical school or single thinker. Certainly it is misleading to see a direct link between Bohr and Kierkegaard, as some have claimed. Bohr was aware of Kierkegaard's work, and we know that as a student he had read at least *Stages on Life's Way*. But this fact is hardly remarkable, for Kierkegaard is regarded by Danes as a master of Danish prose, and virtually every educated Dane has read something of Kierkegaard's work. We know that in spite of his enthusiasm for *Stages on Life's Way*, Bohr prepared some notes for discussion with his brother against Kierkegaard's general position. However, the attempt to find a parallel between Kierkegaard's "leap of faith" and Bohr's conception of "quantum jumps" is without a shred of historical substantiation and must be regarded as entirely fanciful.

It seems safest to conclude that Bohr's admiration was restricted to Kierkegaard as an artist rather than as a thinker, for the adjectives he applies to *Stages on Life's Way* are "one of the most *delightful* things I have ever read ... so *nice* a book", hardly adjectives one would use to refer to an intellectual appreciation of Kierkegaard' ideas.[27] Nowhere in any surviving manuscript does Bohr ever refer to Kierkegaard as a source of inspiration for the ideas of complementarity. Although there are common themes, very probably transmitted via Høffding, it seems that the maximum possible Kierkegaardian influence is that suggested by Holton:

> Now it would be as absurd as it is unnecessary to try to demonstrate that Kierkegaard's conceptions were directly and in detail translated by Bohr from their theological and philosophical context to a physical context. Of course they were not. All one should do is permit oneself the open-minded experience of reading Høffding and Kierkegaard through the eyes of a person who is primarily a physicist—struggling, as Bohr was, first with his 1912–13 work on atomic models, and again in 1927, to "discover a certain coherence in the new ideas" while pondering the conflicting, paradoxical, irresolvable demands of classical physics and quantum physics which were the near despair of most physicists of the time.[28]

Any attempt to link the two thinkers further stretches the historical evidence beyond the breaking point.

Professor Jammer has speculated on the possible influence of the French philosophers Charles Renouvier and Emile Boutreaux. There is absolutely no mention of these thinkers in any surviving writings of Bohr (virtually all of which have been saved), so establishing a direct influence is out of the question. Høffding was aware of these thinkers and may have passed on their ideas, but it would seem far safer to argue that at best the ideas of Renouvier and Boutreaux were simply indicative of the winds of doctrine of the time. Both men stressed the theme that natural law approximates reality and, if taken literally, leads to a falsification of experience. Furthermore, both philosophers defended the ultimateness of experience as the fundamental ontological category, that which really can be said to be. According to this view, the description of experience in terms of subjective and objective aspects refers to abstractions generated by the artificial imposition of arbitrary conceptual forms on the unity of experience. Though arbitrary from the point of view that other descriptive concepts might equally have been employed by a different mode of description, it is necessary that some such abstraction or another is invoked in order to be able to describe experience at all. As we shall see, these themes are important in complementarity.

In his two published testimonials to Høffding (on the occasion of Høffding's eighty-fifth birthday and again at his death) Bohr stressed how much he had learned from his philosophical mentor, but in no case does he ever state exactly what he believed Høffding had imparted to him.[29] Thus we cannot say with any certainty whether Høffding had taken up these ideas or whether they influenced the young Bohr. However, it is certainly possible, though we cannot say how likely, that either in discussions with Høffding at the Bohr home or with Høffding's students in the Ekliptika Circle Bohr became acquainted with the idea that experience is of a piece and that as soon as any term used to describe experience is given an unambiguous meaning (as is necessary in science), an arbitrary separation is necessarily introduced between experiencing subject and experienced object. Furthermore, from what we do know of Bohr's early epistemological speculations, as we shall see shortly, he seems to have regarded these ideas as he key to solving the problems which philosophers had treated in "an irrelevant manner".

Thus two themes come together in the influence from his father and his mentor. First there is the recognition that philosophy of science is concerned with the concepts through which experience is described and one must become aware of the limitations of the applicability of such concepts to specific phenomena. Second, there is the notion that the description of experience in terms of a subject which experiences an object imposes an arbitrary dichotomy onto experience which though necessary to make possible an objective (unambiguous) description of experience, generates a descriptive abstraction in which the wholeness or continuity of experience is belied.

There is one strong clue to guide our analysis of how the young Bohr may have reflected on these themes, for we can say without any doubt that the one philosopher for whom Bohr had strong appreciation, in the sense of dramatic agreement, was William James. One of the members of the Ekliptika Circle was Bohr's friend, Edgar Rubin, later to become a psychologist; apparently Rubin recommended that Bohr read James. Bohr's recollection from the interview the day before he died is worth quoting:

> I was a close friend of Rubin, and, therefore, I read actually the work of William James. William James is really wonderful in the way he makes it clear—I think I read the book, or a paragraph called ... "The Stream of Thoughts", where he in a most clear manner shows that it is quite impossible to analyze things in terms of—I don't know what one calls them, not atoms. I mean simply if you have some things ... they are so connected that if you try to separate them from each other it has nothing to do with the actual situation. ... I know something about William James ... I thought he was most wonderful.[30]

Unfortunately, there was no chance to follow up these comments for more detail, and as it occurs, there has been some controversy over dating the time of Bohr's first exposure to James. In 1962 Bohr himself remembered definitely having read James before 1912, *i.e.* long before complementarity appeared in 1927. However, Rosenfeld remembers distinctly that he and Bohr first encountered the ideas of William James only in 1932, well after complementarity had been worked out in its essentials.[31] This is a question of some importance, for in many ways complementarity closely parallels James' philosophical problems in the descriptive task of psychology. Furthermore, if in fact James' outlook expresses the direction from which Bohr's thought comes, then the imputation that complementarity is of Kantian origin would seem to be incorrect, for in the very chapter to which Bohr refers, "The Stream of Consciousness", from *The Principles of Psychology*, James is arguing vehemently against the Kantian approach to the description of experience. I shall postpone discussion of the question of a Kantian influence until Chapter Seven, when we have become better acquainted with complementarity itself and are prepared to evaluate its epistemological status. For the present, however, it might be noted that Bohr certainly *thought* his position was in agreement with James, and by implication opposed to Kant's, which Bohr viewed as an *a priori* rationalist attempt to justify the universality and necessity of the framework of classical physics, the very view complementarity was designed to replace. For complementarity, as for

James, experience is ultimate; the "subject" and "object" which "interact" in the production of experience are terms that refer to abstractions invoked for *describing* experience. (James also admired Renouvier who had parallel views.) If, then, the appearance of a Kantian foundation of complementarity is an illusion (as I shall argue is the case in Chapter Seven) and Bohr would find his true philosophical cousin in William James, the question of James' influence on the young Bohr is not merely of historical importance, for it may well help us to understand what Bohr intends to say.

Professor Holton has put to the would-be interpreter of Bohr the dilemma of choosing between either Bohr's remembrance or Rosenfeld's, a decision for which it might seem we have equally divided evidence.[32] However, there is some grounds for tilting the balance towards Bohr's recollection. Høffding had visited William James in America in 1904, during which time Bohr was Høffding's immediate student. The two philosophers found themselves in great sympathy with respect to the common rejection of absolutist metaphysics and their conviction that any single description of experience robs it of its essential nature. Høffding apparently acquainted James with the work of Kierkegaard, whom both regarded as a common ally against the idealist metaphysicians. Clearly their meeting left deep impressions on both men. That Høffding would have then returned to Denmark and made no mention of James to his circle of students and friends, including the Bohrs, both father and son, seems completely incredible. Thus it seems highly probable that while Rosenfeld may be right about when Bohr *read* James, *i.e.* not until 1932, Bohr could well have been reasonably acquainted with James' ideas from approximately 1904 on.

However, the case can be made even stronger, for Bohr explicitly links his remembrance of reading James with Edgar Rubin, a psychology student and member of the Ekliptika Circle. The reputation of James' *Principles of Psychology* was so immense at this time (especially in Europe) that surely at least Rubin must have been acquainted with it long before 1932. We know that at this time in the discussions of the Ekliptika Circle and on his own the young Bohr was struggling to come to grips with the problem of describing the contents of psychological processes, an interest common to both his father and Høffding as well. It seems extraordinary to suppose that both Rubin ad Høffding would have listened to Bohr's meditations on these themes and not noticed the obvious connection with James' problems in *The Principles of Psychology*. Furthermore, the one central problem in both psychology and philosophy to which Bohr returned time and again is the very same issue which was central in James' philosophical development, the issue of freedom and determinism. At the time this issue was generally treated in the context of the mechanism versus vitalism debate, and thus of central interest to Bohr's father and Høffding. Thus again it seems incredible that this issue would have been "on the table" in the discussions between Bohr's father and Høffding as well as in those of the Ekliptika Circle, and yet James' views were not considered.

One possible way of harmonizing Bohr's and Rosenfeld's recollections would be to argue that Bohr became acquainted with James' philosophical *psychology* as early as 1904, but that he remained unaware of James as a *pragmatist* until the later date of 1932. This interpretation is reinforced by the fact that when Rosenfeld brings up James' name, he does so in connection with the claim that complementarity is essentially of a *pragmatist* character, whereas Bohr's recollection explicitly refers to James' *psychological* work. Indeed it seems reasonable that in 1904 this would have been the only James whom Bohr would have been likely to have read and that the later pragmatic writings are sufficiently distinct that when Bohr became aware of these essays at a later date, it was like encountering a new figure altogether. Thus I would conclude that if any philosopher other than Høffding is to be given credit for having shaped complementarity, that credit belongs to William James. Through James the influence of Renouvier and Boutreaux, both of whom James greatly admired, could be said to have had a third hand effect on Bohr's thought, but it is likely that Bohr himself was never aware of this.

4. Early Epistemological Work

As mentioned in the previous section, Bohr's earliest philosophical interests focused on the question of how to describe phenomena, including psychological processes and the relation of this to the issue of free will. Following these interests the young Bohr contemplated a work in epistemology which would analogize the description of psychological states to the problem of mapping multivalued functions of complex variables developed by G.F.B. Riemann. Unfortunately, the components of this analogy are extremely obscure in Bohr's sole surviving reference; hence any reconstruction of his thoughts is necessarily speculative.

In the treatment of this problem, Riemann proposed that for each single value of an independent variable, the different variables are mapped onto different planes. Bohr analogized each descriptive term in a language to such a function having many "values" or, in the analogy what he called "planes of objectivity". Thus a single term will have different descriptive functions depending on what is considered to be the "object" of the description, or in other words, on which plane of objectivity we are giving a description. For example, if I report "seeing a red apple" the "object" of my description of this phenomenon may be taken to be a fruit, or the rays of red light entering my eye, or the sensory stimulus thereby produced, or the psychological event of having the "idea", "impression", or "representation" of the apple. Normally, the context of the description is sufficient to establish whether the "plane of objectivity" is intended to be aesthetic, physical, physiological, or psychological. But when we venture into unfamiliar areas of experience, the use of familiar descriptive terms may become ambiguous owing to an unclarity in which plane of objectivity is intended. In order to avoid the ambiguity involved in knowing which value of the multivalued function is being considered, Riemann mapped these different variables onto different planes, each plane representing a different set of values of a single *valued* function for each value of the independent variable. Bohr's analogy intended to show that the philosophical problems surrounding the free-will question and the description of psychological processes result from an incautious use of descriptive terms which become systematically "ambiguous" when the experiencing subject attempts to describe as an object his own experiencing activity. Such problems could be avoided by keeping distinct "planes of objectivity" for the different uses of the descriptive concept. Thus it seems that he would have argued that in the report, "I did action A by an exercise of free-will", the "I" to which we refer is what Kant called the "transcendental ego" which is always subject and cannot be taken as object in a sense which would give definable meanings to a cause and effect deterministic account. Thus we generate a conflict between two apparently rival descriptions: that of the causal deterministic account of the phenomenon of an individual's choice and that of the individual who reports an immediate experience of freedom in choosing to do an act. But this apparent conflict is an illusion created by the fact that we fail to notice that the object of description is different in the two accounts owing to a shift in the plane of objectivity on which each description is offered.

Since our immediate task is not to solve the problem of free-will, the instructiveness of this speculative analogy need not concern us. What is of interest, however, is the parallel between this attempt to treat the question of the use of descriptive concepts and Bohr's later treatment of the same issue in complementarity. Apparently Bohr was not really aware of the striking similarity which his proposed "solution" to the problem of free-will bears to that of Kant's, for he undoubtedly considered it a totally original contribution.

It is very likely that Bohr discussed this analogy at considerable length with his brother Harald, for the latter did his doctoral thesis on Riemann. This fact helps us date these speculations as having occurred before 1910, for in June of that year Niels wrote Harald, in a merry mood, that he did not know which of several fortunate events had caused him his present positive emotional state. Interestingly he chose the following words: "I must confess I don't know if I am most happy over your appointment, over the good behavior of my electrons at the moment, or over this portfolio; probably the only answer is that emotions [*Følelserne*] like cognitions [*Erkendelsen*] must be arranged in planes that cannot be compared."[33] The reference to planes could hardly be expected to make much sense if the brothers had not discussed the Riemannian analogy at some earlier date.

I will return to this analogy in Chapter Six in discussing the application of complementarity to psychological descriptions, where we can employ it instructively. At present I intend only to provide further evidence that at this early date Bohr was already deeply concerned with the relation of subject and object in experience. Furthermore, he did not focus on this relationship in the Kantian manner of asking what is the nature of the experiencing subject such that the object of experience has the form it presents. Instead, he saw this issue entirely in the context of learning how to use descriptive terms in a way which avoids ambiguity.

One final certain philosophical influence on Bohr deserves mention in this connection. It is what Bohr held to be without a doubt his favorite exposition of philosophical problems, an unfinished novel by Poul Martin Møller, *The Adventures of a Danish Student*, a work which Bohr, like all young Danes, had certainly read in his school days. Møller was a professor of philosophy at The University of Copenhagen from 1831–1836, during which Kierkegaard had been his student and admirer.[34] This light-hearted book is widely attested to as Bohr's favorite work of literature, for he regarded it as an exemplary lesson in epistemology. In Bohr's words, it "gives a remarkably vivid and suggestive account of the interplay between the various aspects of our position, illuminated by discussions within a circle of students with different characters and divergent attitudes to life".[35] Rosenfeld records that "Every one of those who came into closer contact with Bohr at the Institute, as soon as he showed himself sufficiently proficient in the Danish language, was acquainted with the little book: it was part of his initiation."[36] According to Rosenfeld, this "delightfully humorous illustration of Hegelian dialectics ... would one day start a train of thought leading to the elucidation of the most fundamental aspects of atomic theory and the renovation of philosophy of science".[37] While this claim may well be exorbitant, it does establish the degree to which those who knew Bohr well considered this novel as influential in the development of complementarity.

The point which Bohr used Møller's novel to teach the novitiates at his Institute parallels the Kantian distinction between the transcendental ego and the empirical ego, the importance of which already appeared in Bohr's Riemannian analogy. Again Bohr reveals that he sees this distinction as one referring to the "problem of the unambiguous communication of experience", which he believed to be the heart of the problem of knowledge.[38] The character who interested him most, called "The Licentiate", is one who "is addicted to remote philosophical meditations detrimental to his social abilities".[39] Bohr quoted his favorite passage in one of his essays:

> I get to think of my own thoughts of the situation in which I find myself. I even think that I think of it, and divide myself into an infinite retrogressive sequence of "I's" who consider each other. I do not know at which "I" to stop as the actual, and the moment I stop at one, there is indeed an "I" again which stops at it. I become confused and feel a dizziness as if I were looking down into a bottomless abyss ....[40]

Clearly this "dizziness" was one which the young Bohr had felt, but unlike The Licentiate, it was one which he determined to overcome. His first answer was given in terms of the Riemannian analogy and its "planes of objectivity". As we shall see, in a very significant sense, the framework of complementarity can be regarded as a more mature and sophisticated answer to the same problem.

The reference to the phrase "bottomless abyss" in the the previous quotation calls up striking associations, for it was one of Bohr's favorite phrases. In response to Kierkegaard's claim that he found himself with 70,000 leagues of water beneath him, Bohr would add, "with a twinkle in his eyes: 'It is much worse—we are suspended over a bottomless pit, caught in our own words' ".[41] Again, his favorite quotation from poetry was from Schiller's *Sayings of Confucius: "Nur die Fulle führt Klarheit/ Und im Abgrund wohnt die Wahrheit"* (Only wholeness leads to clarity/ and Truth lies in the Abyss).[42] This characteristic of always penetrating to the ultimate foundation for any description and the refusal to accept a description which denied some part of the whole phenomenon may well be taken as the central vector in Bohr's intellectual life. It is striking, therefore, that the same sentiment is expressed by none other than the founder of atomism, Democritus, in whose footsteps Bohr followed some two-thousand years later. That fragment is, "We know nothing of reality; for truth lies in an abyss."[43]

CHAPTER THREE

QUANTUM THEORY AND THE DESCRIPTION OF NATURE

Without exaggeration we may say that the framework of complementarity which Bohr proposed in 1927 was the result of work on atomic theory which he began as early as his doctoral dissertation in 1911. Other analyses of Bohr's philosophy have tended to treat complementarity as though it simply appeared in 1927 as an interpretation of the finished quantum theory of 1925–26. But such treatments ignore the roots of Bohr's conceptual revolution in his original atomic theory of 1913, the so-called "old quantum theory". Because they miss the path that Bohr actually followed to his new framework, they cannot understand complementarity as its author understood it, typically failing to appreciate the crucial role of "the quantum postulate" which always figures centrally in Bohr's presentations.

My method, however, will be to approach complementarity as Bohr approached it. Consequently in this chapter we will follow Bohr's intellectual course from 1911 to September of 1927, when he presented his first statement of complementarity at Como. Conveniently this journey can be marked off by two revolutionary insights on Bohr's part, one at its beginning, leading to his renowned model of the atom, and the other at its end, leading to complementarity. Both conceptual breakthroughs allowed Bohr to see the problem in a new perspective from which he could envision his proposed solution to the relevant difficulties. Of course these two breakthroughs are intimately related: the original one defined a set of problems to which the final proposed a solution. The period in between includes the dramatic story of the struggle to produce a coherent, consistent description of the atom as a physical system. This chapter will refer to that story only *en passant* for our central focus is understanding complementarity, not a history of quantum theory. We shall follow Bohr's odyssey only insofar as it marks the progress of his thought towards his first formulation of complementarity.

1. Bohr's First Great Insight: The Quantum Postulate

Bohr understood the fundamental task of atomic physics to be accounting for the properties of the chemical elements in terms of atomic structures. Of course the term "atom" in its original sense referred to an elementary constituent of material bodies. However, well before Bohr's entry into physics it was clear that the "atoms" which were the fundamental units of the chemical elements were not in fact elementary in the sense of having no internal structure. The various properties of the chemical elements exhibited patterns that called out for an explanation, and that explanation was generally understood to require describing the structure of the chemical atoms. These "atoms" were thus conceived as physical systems composed of yet more elementary constituents which existed in a structure that was capable of being described by the equations of motion of classical mechanics. Insofar as the chemical atom was a physical system composed of such constituents, the presumption was that the behavior of such a system could be described within the framework of classical mechanical and electrodynamical laws.

Because Bohr's revolutionary atomic theory required a dramatic break with classical theory, it is necessary to pause briefly to review how classical physics described the behavior of physical systems. As an empirical science, mechanics must of course develop a set of concepts for describing what will be observed in specific circumstances. Since motion is change of position through time, for a science of mechanics what needs to be observed is only the positions of bodies at instants of time, and the subsequent change of these positions through time relative to some reference system. Since the term "body" refers to those objects whose behavior is described in mechanics, it follows by definition that bodies must possess the properties of position at each instant throughout a temporal duration, as well as the ability to change those positions. A "physical system" may then be defined as any group (including possibly one) of such bodies.

Mechanically the simplest possible system would be a single body system whose position could be represented as at rest or moving in a straight line at a constant velocity (uniform rectilinear motion) relative to some reference system. Newtonian mechanics stipulates that an acceptable description of such phenomena needs no cause external to the system to explain its continuance in such a state. As long as a system remains in such a state, it may be described as "closed" or "isolated", *i.e.* as free from any external causal effect on its mechanical behavior.

But of course mechanics cannot provide descriptions of only single body systems whose motion does not change. Real physical systems consist of more than one body and of bodies not at rest or in uniform rectilinear motion. The cause of departures from such motions is defined as a "force". Thus an adequate mechanical theory must provide a way for describing the causal effect on the motion of a body produced by a force acting on that body. Any phenomenon in in which such a force manifests its effect on the motion of a body may be called an "interaction". Thus a complete science of mechanics must include a way of describing interactions which permits determining their effect on the motion of bodies.

An interaction can occur by contact force as in a collision between two impenetrable bodies, or it might be represented as "acting at a distance" as is the gravitational force of one body on another. Thus not all interactions in which physical systems may be involved are represented as occurring directly between material bodies. In the development of atomic physics what was of particular concern was the interaction between atomic systems and light or electromagnetic radiation. The classical framework was able to account for most of the phenomena exhibited by light radiation by representing these phenomena as the consequence of the behavior of the "electromagnetic field". Though not visible or impenetrable like material bodies, fields are defined as existing over a spatial region and manifesting a force at each point in that region. Thus in classical electrodynamics light and other forms of electromagnetic radiation can be theoretically represented as a wave disturbance moving in an electromagnetic field, commonly analogized to the motion of waves across the surface of a liquid. Just as the mechanical behavior of material bodies is described by tracing the spatio–temporal trajectories of bodies, so the electrodynamic behavior of the field (as manifested in its interactions with material bodies) can be described in terms of the propagation of waves through the field.

Although any two interacting bodies may be considered as separate interacting systems which are not isolated, together they might equally well be considered to be components of a single physical system in which all interactions are internal to that system. In this latter case, such a system of interacting components may be "closed" or isolated from interaction with bodies external to the system. Classical mechanics defined the concept of the "mechanical state of a closed system" at each instant of time in terms of theoretical parameters, the "classical state parameters", which would allow one to define the state of the system at any future instant as long as it remained closed. To define its mechanical state, each body in the closed system could be characterized in terms of two parameters, position and momentum, for each instant in time. A "particle" can be defined as any body possessing essentially properties corresponding to these state parameters.

A mechanical description of the interactions between matter and radiation could be expressed in terms of an exchange of energy between bodies and the field. A "wave" propagating through the field is the way which classical electrodynamics represents the transmission of light or other electromagnetic energy across space over an interval of time. Such a wave disturbance moves through the field at the speed of light, and is defined in terms of its energy and frequency, *i.e.* the number of cycles per unit of time. Thus within the classical framework, the concepts of "particle" and "wave" refer to the theoretical representations through which one describes the behavior of matter and radiation.

As long as the system remains closed, if we know the state parameters for each component in the system at some point in time, the "initial conditions", then the laws of mechanics, when applied to the state of the system, permit one to define the state of the system at any future time. In the interactions between components of a closed system these components will exchange momentum and energy according to well defined classical principles, but as long as the system remains closed, the total amount of momentum and energy in the system as a whole remains constant, or is "conserved". The conservation principles for momentum and energy are thus what make possible defining the future state of a system by applying classical principles to its state at some initial moment.

A theory of atomic structure must provide a way to represent the physical behavior of the constituents of the atom as a physical system so as to account for the properties of the chemical elements; *i.e.* it must provide an "atomic model". There seemed good empirical reason to believe that such a model must include the electron, discovered by J.J. Thomson in 1897. The thinking which led to Thomson's discovery presupposed essentially that these elementary constituents of matter are themselves material bodies. Thus the electron could be described as a particle carrying a unit electric charge. However, all attempts to build atomic models on the assumption that the atomic constituents behaved according to classical laws ran into conflict with the empirically established properties of the chemical atoms.

Beginning with his master's thesis of 1909 and continuing through his doctoral dissertation of 1911, Bohr analyzed the then existing electron theory of metal elements in a systematic attempt to trace electrical and other properties of metals to the presence of "free electrons", *i.e.* electrons not treated as essential parts of the atomic system. Although Bohr certainly had no doubt that metallic properties could be accounted for by atomic structure, his dissertation marked his precocious insight into the limitations of applying classical mechanics to such systems.

In the electron theory of the Dutch physicist H.A. Lorentz, the interaction between electrons unattached to atomic systems and the molecules of a metal could be described by the classical treatment of collisions between bodies. In his analysis of the theory, Bohr observes that presupposing the applicability of classical mechanics is "not a priori self-evident, for one must assume that there are forces in nature of a kind completely different from the usual mechanical sort".[1] Here we see that at the very start of his career Bohr had no commitment to any *a priori* foundation for the classical conceptual framework for describing nature. Indeed, Bohr rejects the classical viewpoint on explicitly empirical grounds: "there are ... many properties of bodies impossible to explain if one assumes that the forces which act within the individual molecules ... are mechanical also".[2] In fact this orientation hardly marks a by-the-way attitude on Bohr's part, for as Kuhn and Heilbron have observed, Bohr's "conviction of the ultimate incompetence of 'the ordinary mechanics' in atomic theory sets the tone of the thesis".[3]

As we saw in the previous chapter, Bohr's method required searching for a breakthrough by cumulatively compiling the then known difficulties which any theory had to confront. This approach was already evidenced in his dissertation where he tries to show that the range of unexplained phenomena which the electron theory would have to surmount requires more than merely modifying "special assumptions" about how the electrons interact with molecules. Instead it requires abandoning any mechanical description which stays within the classical framework. Furthermore, by compiling existing difficulties to get at the foundations of the theory, Bohr recognized that J.J. Thomson's approach, which tried to describe the interaction of the electromagnetic field with electrons bound to atomic systems through classical means, would lead to absurd results.

Thus it comes as no surprise that immediately after completing his dissertation, when Bohr went to Cambridge to interest Thomson in his work, the latter was hardly patient at lending an ear to Bohr's criticisms. However, Thomson's work was not unfruitful, for it is likely that Bohr's consideration of Thomson's atomic model led to his early conviction that the break with classical mechanics would involve not *free* electrons, but electrons considered as part of the physical system of the chemical atom, *bound* electrons. In fact this recognition marks the first suggestion that explaining the properties of the chemical elements would require a non-classical description of atomic structure. This awareness is the conceptual bridge between Bohr's student interest in the electron theory and his ultimate accomplishments in atomic theory that flourished when he moved to Manchester. What is significant for our purposes is that what propelled Bohr across this bridge was his conviction that classical mechanics must break down in the atomic domain. Thus he began to search for the point of most tension within classical attempts at describing atomic systems, a point which he hoped would localize the exact region where the necessary reconstruction must occur.

In the spring of 1912 when Bohr moved to Manchester to work with Rutherford, his interest in the electron theory gradually changed to a new interest in atomic models. One year earlier, by analyzing the interactions of atomic systems with bombarding particles, Rutherford had determined experimentally that such systems could be described as having a massive positively charged nucleus surrounded by negatively charged electrons. But the mechanics necessary to describe such systems remained completely unclear. Rutherford, who was basically an experimentalist, was not inclined to pursue the logic of his proposed model in any detail, but it was for just such a task that the young Danish newcomer was perfectly prepared.

Bohr approached this task from the point of view of the crucial atomic property of stability. The chemical atom was introduced as that which endured through chemical alterations. From a mechanical point of view this fact meant that if the chemical atom was to be described as a mechanical structure of fundamental particles, then this system must be mechanically stable. Over and over again in his early lectures Bohr emphasized that this property is crucial to building a successful atomic model.[4] However, all attempts to use classical mechanics to determine the mechanical structure of the chemical atom produced a physical system which was mechanically unstable.

In order to describe the interaction between matter and the electromagnetic field, the classical framework represented particles as possibly carrying either a positive or negative electric charge. Rutherford's discovery of the positively charged nucleus implied that the negatively charged electrons in the atomic system would be drawn in toward the nucleus by the force of electrical attraction. To counteract this tendency of the atomic system to collapse, it was imagined that the electrons could orbit the nucleus in a way resembling the planetary orbits already well described by classical mechanics. In describing classically a physical system consisting of a particle moving in an electromagnetic field, when the charge of the particle is determined, the nature of its interaction with the electromagnetic field can be predicted by the laws of electrodynamics. A charged particle moving in a cyclical path may be considered an "oscillator" defined by a specific "frequency" or number of times the particle completes the cyclical path per unit of time. Such a charged oscillator will impart energy to the electromagnetic field, and this energy can be represented as a wave moving through the field, the frequency of which is proportional to the frequency of the oscillator. However, this description implied that the orbiting charged electron would behave as an oscillator, giving up energy to the surrounding electromagnetic field. As the electron lost energy its distance from the nucleus would shorten. Thus it would spiral into the nucleus, causing the atomic system to collapse. To a young physicist like Bohr in 1912 these failures pointed to the need for breaking with classical mechanics in order to describe a successful atomic model.

Already in his dissertation Bohr had considered the significance of what was then the greatest apparent break with classical mechanics: Max Planck's quantum explanation of "black-body radiation". Planck had proposed his hypothesis to describe the nature of the electromagnetic radiation given off by an idealized thermally excited body (a "black body"). According to classical theory, if a black body is made up of atoms containing oscillating charged particles, electrons, such a body will produce electromagnetic radiation that will have a specific amount of energy for each frequency. However, all attempts to fit this theoretical description to the observed electromagnetic radiation actually emitted by radiating bodies had failed because they were based on the classical assumption that the charged particles may be oscillating at any possible frequency. Planck proposed that the average energy of the atomic oscillators composing the material body is "quantized", *i.e.*, it occurs at only certain discrete values determined by integral multiples of a fundamental physical constant, now known as "Planck's constant".[5] Many physicists attempting to build atomic models recognized that there may be some connection between Planck's hypothesis and the atomic constitution of radiating bodies. Yet the connection was not clear, for most physicists, including Thomson, then understood Planck's break with the classical assumptions as a heuristic device for furthering research by making the apparently false assumption that a non-classical quantization occurs at the atomic level. To those physicists, Planck's constant was "derived" and in the end would be reduced to the continuous parameters of the classical framework.

But by 1912 Bohr was probably already convinced that Planck's hypothesis had uncovered a fundamental fact of nature that could not be analyzed along classical lines. The extent to which Bohr connected this fact with atomic structure is uncertain before 1912, but his insight in Manchester in June of that year demonstrates that he recognized that Planck's constant was the key to the non-classical principles which, as he had already foreseen in his dissertation, must apply within the atomic system.

Throughout the winter of 1911–12 Bohr studied Planck's quantum hypothesis, about which he wrote, "I am very enthusiastic ... but I am not sure if this is due to my ignorance."[6] However, by June of 1912 his interest seemed to have turned towards atomic models, for he wrote his brother Harald that he believed Rutherford's model "seems to be quite a bit more solidly based than anything we've had before".[7] Only one week later he again wrote Harald, this time with barely concealed excitement:

> It could be that I've found out a little bit about the structure of atoms. ... If I am right, it would not be an indication of the nature of a possibility* [the asterisk leads to a marginal note: "i.e. impossibility"] (like J.J. Thomson's theory) but perhaps *a little piece of the reality.*

The marginal note shows that Bohr did not regard his insight as simply ruling out a possible classical description, as had Thomson's model (an "impossibility"), for he had already realized this well before. He clearly regarded his breakthrough as a move towards understanding what was really taking place in atomic systems: "a little piece of the reality". Thus in this first written reference to his revolutionary insight, Bohr does not present his discovery as a heuristic device he has decided to employ. Instead it is clear that his discovery comes from reading Planck's hypothesis as revealing a fundamental fact about the structure of real atoms.

To account for the empirically established facts about atomic systems, Bohr's revolutionary theory presented a bold departure from all previous considerations. It describes the atomic system as characterized by electrons orbiting the central nucleus at certain specific quantized frequencies. Classical theory implied that such oscillating electrons would emit radiation determined by its frequency of revolution about the nucleus, thereby giving up energy to the surrounding electromagnetic field. But in June of 1912 Bohr had the idea to represent the atomic system as existing in a "stationary state" in which it did not radiate at all. This decision to adopt as a radical hypothesis what he later called "the quantum postulate" entailed assuming that the atomic system exists in only such a discrete state characterized by a quantized value of energy in which, contrary to classical principles, the system is stable. Although at the time (June 1912) Bohr had not seen his way to the exact manner in which Planck's constant was to enter the theory, he did recognize that the break with classical theory would involve such a quantized "stationary state". Only some six to seven months *after* his basic idea of making the atomic system exist in such a stationary state by extra-classical *fiat*, did Bohr develop his theory of atomic structure to include a whole *series* of such states between which the atom undergoes *discontinuous* transitions. At this time Bohr became aware of the fact that electromagnetic energy emitted by the simplest atomic system, that of the hydrogen atom, occurred at specific frequencies which exhibited a numerical relationship to one another given by the "Balmer formula". In postulating that this radiation was emitted by the atom undergoing a transition from one stationary state to another, Bohr's theory gained the impressive advantage of being able to derive this empirically established formula, and thus to explain the line spectra exhibited by radiating atoms.

It is sometimes thought that Bohr's revolution was based on the postulate that electrons must orbit the central nucleus at distinct radii from which they "jump" discontinuously to other radii, thereby emitting discrete quanta of energy. But Bohr first had the idea that the atom exists in a non-classical stationary state, and only later got the idea that radiation occurs when the system "jumps" (*i.e.*, changes state discontinuously) from one stationary state to another. The emphasis on quantum jumps and Bohr's success in describing radiation phenomena observed in atomic spectra tends to obscure the fact that the primary purpose of his theory was to account for atomic *stability* through the quantum postulate. Because the idea to describe the atom as having many possible stationary states came only half a year after his adoption of the quantum postulate, the notion of "quantum jumps" could not have been what led Bohr to his theory, nor was he enticed by any fascination with discontinuity *per se* as has been suggested.[9] His concern from first to last was accounting for atomic stability, the central property which any model had to explain. The apparently classical nature of the orbiting electron model may be deceiving, for of course classical electrodynamics implied such a model would be radically *unstable*. Bohr always thought of the heart of his theory as the concept of "stationary states" in which, by non-classical *postulate*, the system is stable. The now well-known "electron orbits" were simply a pseudo-classical means of representing these stationary states. Though the degree of seriousness with which Bohr took this model waxed and waned throughout this period, certainly he was only rarely tempted to take the orbiting electron as a literal spatio–temporal description of the physical situation inside the atomic system. Furthermore, he was repeatedly alarmed by the tendency of many other physicists to do so.[10] Unlike the classical concept of mechanical state, which by definition changes continuously, Bohr's concept of the stationary states in which the atomic system may exist implied that change to another state is, by definition, discontinuous. Thus the quantum postulate not only is the basis for Bohr's model of the atomic system, but also entails abandoning the classical presupposition that physical systems change state continuously. If atoms exist in only specific discrete quantized states, the change from one state to another cannot take place continuously, for that would require that the atomic system moves through states having every possible value of energy between the two quantized stationary states. But such states are explicitly forbidden by Bohr's quantum postulate.

2. The Development of the Original Quantum Theory

Bohr's adoption of the quantum postulate was completely an *ad hoc* addition to the classical framework with which it sharply conflicted. Thus it was difficult to understand just how Bohr's theory related to the physical processes inside the atomic system. At this time there was no quantum "theory" in the strict sense of a consistent formalism, for once one took account of the quantum exceptions to the classical framework, the remainder of any calculation was forced to use the very classical principles contradicted by the quantum postulate. A quantum theory of the atom needed a single set of consistent principles for describing atomic states and interactions. Thus the importance of Bohr's original atomic theory was that it precipitated and influenced the search for a true atomic theory which would overcome the inconsistencies inherent in this preliminary attempt to combine both classical mechanics and the quantum postulate.

From 1913 when Bohr published his atomic theory until 1925 when Heisenberg and Schrödinger succeeded in formulating consistent theories, Bohr relentlessly followed his general program of extending his early conception of atomic systems to explain all experimentally observed properties of the chemical atom. However, because of its internal inconsistencies he knew from the start that his theory could not provide the ultimately satisfactory description of atomic systems. This strange mixture of classical and quantum ideas had puzzled Rutherford from the beginning, and it remained a primary source of criticism for those who continued to regard Bohr's theory as solely a heuristic device for predicting a variety of phenomena, not a description of what really goes on inside atoms. Thus many physicists expected that once a consistent theory of atomic structure was established, the apparent discontinuities of Bohr's stationary states would be reduced to the continuities of the classical framework. For these physicists the alleged quantum "postulate" would be seen to be strictly false, adopted because of expediency, but abandoned by the completed theory.

Bohr himself cautioned against taking his model as a literal picture of the atomic system not because he suspected the quantum postulate was not a true fact of nature or because he wanted a return to some classical representation, but rather because he realized that a single consistent formalism had not been achieved. Throughout this period he stressed that a new framework for describing atomic systems would require "generalizing" the classical framework. As a guide to such a revision of the classical ideas, Bohr focused on the fact that the classical framework provided an adequate description of mechanical interactions involving a large-scale change of state relative to Planck's quantum. Thus he reasoned that when one described interactions where the discontinuities in change of state measured by Planck's quantum could be regarded as negligible relative to the degree of change of state, the predictions of an acceptable mechanics of atomic systems should correspond to those of classical theory. In other words, Bohr recognized that since interactions involving a large-scale (relative to Planck's constant) change of state characterizing a physical system were adequately described classically, the quantum description of such interactions must tend to "correspond" to the classical description.[11]

This so-called "correspondence principle" allowed the theory to be extended in many significant ways. Moreover, it expresses the physical basis for Bohr's claim that complementarity is a "rational generalization" of the classical framework, because the continuity of change of state in classical mechanics could be understood as a "special case" where the discontinuities expressed by the quantum postulate were so relatively small that they could be ignored and the interactions between systems described continuously.

As soon as the successes of Bohr's theory in describing the observed properties of the simplest elements were widely known, other physicists began to refine his treatment and apply his approach to describe the structure of more complex atomic systems. Although a steady stream of solutions for particular problems was forthcoming, for the most part attempts to understand the nature of atomic systems within this inconsistent framework of the so-called "old quantum theory" were disappointing and frustrating.

Thus while those physicists not so deeply concerned with the nature of the atom as a physical system may have been delighted with the progress being made, Bohr obviously became more and more distressed over the inability to see around the obstacles that stood between his early theory and the attempts to formulate a consistent framework for describing atomic systems. For this reason, instead of moving into the role of revolutionary turned defender of the newly established order, Bohr tended to become more and more revolutionary. But his state of mind was obviously divided, for as the knowledge of his theory and its successes grew, he was increasingly called upon to defend the quantum view against attempts to return to the classical representation. Yet in many ways Bohr remained his own theory's most serious critic, seeking clarity in the "depths" of the inconsistencies which the quantum postulate implied.

3. Discontinuity and the Conservation of Energy

From a general survey of Bohr's manuscripts in the years prior to 1927, it appears that uppermost in his mind were three aspects of the classical framework: what he called "space–time description" as that which is determined by observation, the "claims of causality" which enable one to define the state of a system not observed, and the presupposition that physical systems change their state continuously through time. For nearly a decade before he formulated complementarity Bohr applied a great deal of thought to disentangling how these three themes are woven together in the classical framework. His groping towards complementarity was essentially conditioned by his attempts to understand how these three themes were interrelated so that he could determine how the first two would be altered when the third was denied. Thus we need to understand how these themes were woven into the fabric of the classical framework.

Since Bohr had explicitly denied the classical view that the mechanical state of a system changes continuously in his 1913 theory of the atom, the consequences of this introduction of discontinuity weighed heavily in his reflections on how the classical framework would have to be changed to be made consistent with the quantum postulate. This notion of continuity of change of state can be made more explicit through its representation in "phase space". If we consider a system of a single particle moving in a straight line, its state will be completely defined at any instant by two numbers giving the values of the parameters corresponding to its position and momentum. Each pair of numbers defining the system's state at an instant can be represented as a point on a two-dimensional plane or "phase space" in which one co-ordinate axis represents position and the other momentum. For more complex systems more parameters will be needed and the appropriate phase space will have a corresponding number of dimensions. As the system's state changes through time, it will be represented by different points in the plane, thus tracing out a "path" from the point symbolizing its state at some initial moment to its state at any later time. If this change of state is represented as taking place continuously, this path will be a continuous trajectory in phase space.

As Bohr understood classical mechanics, the ideal goal of observation was providing a space–time "picture" or "model" of the motions of the bodies forming a closed system. Bohr called such a description a "space–time description" or a "description using the mode of space–time co-ordination". Clearly in describing the motion of the objects of ordinary human experience such a description could be the result of simple observation. Thus Bohr spoke of space–time descriptions as fulfilling the "ideal of observation". This idiosyncratic way of speaking about the task of mechanics led to his unusual expression of complementarity in its original formulation. Since it did not prove a felicitous way of making his point, the phrase drops out of later presentations of complementarity. But our understanding of how Bohr developed his thought will be facilitated by catching the point in our story where this idea originates.

However, the mode of space–time coordination is not sufficient for an adequate mechanics, for it is necessary to be able to define the mechanical state of systems after they have interacted with another system. As we saw in the first section of this chapter, in order to describe such interactions, mechanics must add to its mode of space–time coordination a way of describing interactions which permits determining their effect on the motions of the interacting bodies. Bohr referred to this aspect of mechanics as the mode of description which applies the "claim of causality". When referring to the conjunction of the two types of description, Bohr spoke of "the causal space–time mode of description which characterizes the classical physical theories".[12]

Within a closed system the component bodies will interact with each other, but since the whole system is closed the total momentum and energy of the system is conserved. Thus if the mechanical state of one component of an interaction within a closed system is known both before and after the interaction, it becomes possible through applying the principles of conservation of momentum and energy to determine the way in which the interaction changed the state of the other components of the system. In this way the conservation principles make possible applying the claim of causality.

But to do this one must start with "initial conditions" determined by observation. Since an observation requires that the observed system has some causal effect on the observing system, either a human sense organ or a piece of laboratory equipment, a physical interaction between the observed system and the observing system must take place. However, this fact implies that the observed system is no longer closed or isolated, as it must be if we are to apply "the claim of causality" to predict its future state.

Of course in the classical situation the observational interaction was generally between a relatively large sized object, such an an apple or a planet, and a human sense organ. Thus the effect of the observing interaction could be considered so negligible that it could be ignored. But even in those cases where it could not be ignored, the "claim of causality" permitted describing the observing interaction such that the causal effect, or "disturbance", produced by the observation could be determined and taken into account in defining the system's state as isolated from the observing interaction. But strictly speaking the state of the *isolated* system which must be defined in order to predict the future mechanical behavior of the system is *not* determined by *observation*, for that is physically impossible, but it is *defined* by applying the "claim of causality". Thus Bohr spoke of the mode of description employing the claim of causality as "symbolizing" the "idealization of definition". Like the phrase "space–time mode of description" this phrase was so obscure that Bohr eventually abandoned using it. But it is another essential clue to how he reached his ideas.

Because systems are represented as changing their states continuously, in the classical framework a physical system exists in a well-defined mechanical state at all times, whether or not it is interacting with another system. Thus at each point in an interaction it is always theoretically possible to define separately the state of each of the interacting systems; consequently it is always possible to distinguish between the observed object and the physical systems used as observing instruments. It is essential to the classical framework that this conclusion is logically defensible on its own premises, for only if it is, can the nature of the "disturbance" created by the observing system's interaction with the object be determined and used to define the state of the system as isolated from interaction. It is this fact which makes it possible for the classical framework to pose as its descriptive ideal the observational determination of the mechanical state of an isolated system. But the presupposition of continuity of change of state necessary to make this descriptive ideal consistently attainable, is often pushed into the background and not acknowledged. Thus it occurs that when that presupposition is explicitly denied, as it was by Bohr, the classical ideal of description becomes physically unattainable and so judged within the classical framework any description based on a theory which denies this presupposition becomes "incomplete" in that it does not satisfy the stipulated goals of a physical description. This fact is what forced Bohr to argue for a change of framework.

If we assume that a system does change its state continuously, its state at the very last instant of an observing interaction is symbolized by a point in phase space which can be made arbitrarily close to the point symbolizing its state the first instant it is isolated. Since on the presuppositions of the classical framework a system exists in a definable classical state at all times, whether it is isolated or interacting with another system, it is always possible to define the system's state at the last instant of interaction, and thus theoretically determine its state as it is "virtually" isolated from interaction.

By denying the classical presupposition that systems change state continuously the quantum postulate imperiled the classical union of the modes of space–time description and the use of conservation principles to satisfy the "claims of causality". As we just noted, the classical descriptive ideal of defining the state of an isolated system through an observation which interacts with it could be a consistent goal only on the presupposition that the change of state of a system can be represented as a continuous path in phase space. Thus establishing a new framework which would harmonize the inconsistency between classical and quantum ideas suggested that the classical ideal of a "causal space–time mode of description" would have to be modified in some serious way. This fact in turn suggested on Bohr's analysis outlined above that one or the other of the two classical modes of description which were combined in the classical framework would have to be abandoned or remolded.

Bohr first turned his attention to the mode of applying the claim of causality through using the conservation principles to determine the change of state caused by an interaction. Since the description of a continuous change of state was made possible by applying the principle of conservation of energy, Bohr was led to consider the possibility that energy was conserved only statistically in the long run over many individual atomic interactions. At each moment in each single atomic interaction with the electromagnetic field, strict energy conservation could not be assumed. Although this suggestion was disconfirmed by experiment, the episode taught Bohr a lesson which took him one step closer to complementarity.

The first event shaping this development was the early suggestion by Einstein that certain electromagnetic phenomena could not be described by representing radiation through the field, but could be described adequately by representing radiation as streams of particles or "light-quanta" called "photons". Such photons were described as having an energy defined by the frequency of the radiation (when described as a wave) multiplied by Planck's constant. This was a strange idea, since it depended on representing electromagnetic radiation as a wave, necessary to give meaning to the concept of "frequency", in order to describe the same light as like a stream of discrete particles. There was no doubt that Einstein intended his proposal as a "heuristic suggestion" not a literal representation of the reality causing radiative phenomena. However, it propounded the very serious question of whether the description of electromagnetic phenomena through the field representation could be considered adequate. The photon hypothesis was successful in accounting for phenomena where electromagnetic radiation interacted with material bodies but could not describe the interference phenomena so well described by the waves-in-a-field representation. Thus no *single* conceptual representation of light was available for describing the full range of phenomena exhibited by electromagnetic radiation.

Although Einstein's photon hypothesis could describe the discontinuous change of state of electromagnetic radiation when an atomic system absorbed or emitted a light-quantum, Bohr stressed that only by the field representation could one describe radiation in free space isolated from its interactions with material bodies as well as the interference phenomena created by the interactions between different emissions of radiation. Hence, Bohr resisted according any "reality" to this particle-like model of radiation. As Klein recalls, "I think he exaggerated, so to say, the non-reality of the photon. ... He always stressed the necessity of looking into the deep abyss between the classical way of looking at things ... while Einstein tried in some way to have a continuous development with the classical ideas."[13] Bohr liked to make his point with dramatic parables; recalling his first meeting with Einstein in which the two discussed the photon hypothesis, Bohr mused, "Does he [Einstein] think that if he could prove they were particles he could induce the German police to enforce a law to make it illegal to use diffraction gratings [devices which produce interference phenomena that can be described only by a wave representation], or, opposite, if he could maintain the wave-picture, would he simply make it illegal to use photo-cells [devices producing phenomena only the photon model could describe adequately]."[14] Heisenberg reports a similar sort of parable: "He [Bohr] once said to me, 'And even if Einstein would send a telegram and would write to me that now he had definite proof that the light quanta exist as reality, even then the telegram could only reach me by the radio waves which are there.' So Bohr in some way saw the reality of waves as a very strong thing."[15] Thus for the very reason that he saw the wave theory as indispensable for the description of radiation in free space, particularly with regard to interference phenomena, Bohr was reluctant to accept the possible reality of photons. Indeed, he always insisted that though the photon theory attempted to represent electromagnetic radiation as particle-like, it could describe phenomena only by defining the frequency of the radiation through representing that same radiation as wave-like.

Typically Bohr believed the ultimate resolution of these difficulties would require understanding the limits of the use of the concepts of "particle" and "wave" for representing radiation in an adequate description of the relevant phenomena. He shows evidence of this state of mind as early as 1914, only a year after his atomic theory, when he wrote his friend C.W. Oseen,

> I am inclined to believe that the problem contains exceedingly great difficulties which can only be overcome by a departure from classical considerations much larger than has hitherto been necessary. ... the possibility for an embracing picture should not be sought in the generality of viewpoints but perhaps rather in the strictest possible limitation of viewpoints.[16]

What is interesting in this letter is that even this early Bohr seems to be searching for a way in which both conceptual representations could be reconciled by finding a limit to the applicability of each. He seems not to have been strongly tempted, as were Einstein, de Broglie, and Schrödinger, to try to do away with one form of representation in favor of the other.

Nevertheless, if Bohr desired to retain the field description of electromagnetic phenomena, then the interactions between radiation and atomic systems could not be accounted for with any strict consistency, for the field, by its classical definition, changed its state continuously, while the atomic systems, by the quantum postulate, changed discontinuously. Because of his reluctance to accept the photon hypothesis, he cast about for some other break with the classical framework which would allow him to retain the field representation of radiation and yet describe the interaction between radiation and atomic systems. For a time he believed that this break would require abandoning strict energy conservation.

The following model may help show why Bohr's reluctance to relinquish the field representation led him to consider abandoning energy conservation. Imagine that water is to be transferred from one vessel to another and that the water can come from the first vessel only in the form of ice cubes, while it can be received by the second vessel only as a liquid in continuous stream. Let us say that at the start of this "interaction" the first vessel contains two ice cubes, each one unit volume of water (a discrete quantum), and that the second vessel is empty. The "system" consisting of both vessels as its interacting components will then have a total volume of two units of water. Now the two systems are made to interact such that at the end of the interaction these two units of water have been transferred to the second vessel. Assuming no evaporation or spillage in the transfer, we might say that from the time before the transfer has begun to the time after it is completed, the total volume of water is "conserved". We can also say quite unambiguously that half way through the interaction there would be a time when each vessel contained one unit of water, or in other words, one ice cube had been transferred from the first vessel to the second. However, there can never be an instant when the first vessel has, say, one and one-half units of water, for water can be removed from it only in whole ice cube units. Nevertheless, there must be an instant when the receiving vessel contains one-half or one and one-half units of water, as well as, for that matter, every possible volume of water between zero and two units, for in this model it is required that the second vessel receive water continuously. Even were the ice cubes melted instantaneously, the ensuing liquid must be poured into the receiving vessel. There would be an instant when we would have removed one full unit volume of water from the first vessel, leaving it with only one unit, but we would have poured only one-half of the now liquid water into the receiving vessel. The sum of the contents of the two vessels would then be only one and one-half units of water. Consequently there would be a time when the total system of the two vessels would contain *less* than the total two units of water with which the system started and must end. Thus we must conclude that the volume of water is not conserved in each single transfer, say at the moment the first vessel has given up one whole unit of water but the second has not yet received that whole unit. But over the whole run of the process, the total volume does remain constant or is "conserved".

This analogy reflects the situation involving the interaction of a discontinuously changing atomic system with a continuously changing electromagnetic field. Here energy would be analogous to the volume of water and the two vessels to the atomic system and the electromagnetic field, respectively. Since the large scale interactions which are adequately described by the classical framework are described as the result of many interactions between individual atoms and the electromagnetic field, to retain both continuity for the field and discontinuity for the atomic system meant that the total transfer of a quantity of energy would be a constant sum only in the long run over the course of many individual interactions between single atoms and the field. If we are to retain the quantum postulate then atomic systems exist only in stationary states and must change discontinuously from one to the other. If we retain the field representation for radiation, then the electromagnetic field with which these atoms interact changes its state only continuously. Consequently, since an individual atom will have given up a whole light-quantum before the field can have absorbed that whole quantum, it would appear that we are logically required to conclude that energy is not strictly conserved in each individual process.

In 1924 Bohr authored a paper together with Hans Kramers, his assistant, and a young American physicist visiting Copenhagen, John Slater. This paper tries to describe the interaction between matter and radiation by assuming that energy is not strictly conserved. Bohr begins by calling attention to the basic dualism between wave and photon descriptions of electromagnetic phenomena. Although any interpretation of the light-quantum hypothesis as describing real entities is conspicuously absent, Bohr admits that "in its most extreme form" this hypothesis "denies the wave constitution of light".[17] On the one hand Bohr's clear preference for retaining the wave description of light is evident in his comment that "the electromagnetic theory of light not only gives a *wonderfully adequate picture* of the propagation of radiation through free space, but also has to a wide extent shown itself adapted for the interpretation of the phenomena connected with the interaction of radiation and matter".[18] On the other hand, Einstein's light-quantum hypothesis merits merely the begrudging comment that

> ... although the great *heuristic* value of this hypothesis is shown by the confirmation of Einstein's predictions concerning the photoelectric phenomenon, still the theory of light-quanta can obviously not be considered as a satisfactory solution to the problem of light propagation. This is clear even from the fact that the radiation frequency ... appearing in the theory is defined by experiments on interference phenomena which apparently demand for their interpretation a *wave constitution* of light.[19]

From these comments it is clear that in 1924 Bohr understood the wave description in the sense of describing the "constitution of light", but the photon hypothesis is merely of "heuristic value". Thus in no case does Bohr want to use the light-quantum idea to describe the transmission of energy from one atomic system to another via the electromagnetic field. Consequently the interaction between matter and radiation must be described in terms of discontinuously changing atomic systems and continuously changing fields. As we have seen the inconsistent mixing of continuous and discontinuous ways of representing the phenomena implies that the interaction cannot be described such that energy is strictly conserved throughout each single interaction between one atomic system and the field. However, since Bohr's commitment to the correspondence principle meant that quantum descriptions must correspond to classical descriptions in cases where the exchange of energy is large, over the long run in the emission and absorption of energy by many atomic systems, the energy is statistically conserved.

We know that Bohr had this idea of abandoning strict energy conservation at least five years earlier, for in 1919 he wrote in an unfinished letter intended for his friend C.G. Darwin,

> ... as regards the wave theory of light I feel inclined to take the often proposed view that the field in free space [*i.e.*, as isolated from interactions with material bodies] is governed by the classical electrodynamic laws and that all difficulties are concentrated on the interaction between electromagnetic forces and matter. Here [*i.e.*, with respect to such interactions] I feel, on the other hand inclined to take the most radical (or rather mystical) views imaginable. On the quantum theory conservation of energy seems to be quite out of the question. ... there is quite apart from the validity in the usual sense, the problem of what becomes of the energy if this principle has to be abandoned.[20]

Apparently Bohr was very unsure of these ideas, for he never sent the letter to Darwin. They remained unsettled in his mind until late 1923 when he began work on the paper with Kramers and Slater, at which time he began to think that abandoning strict energy conservation would enable the theory to describe interactions between radiation and atomic systems.

Bohr recognized more clearly than most physicists that the inconsistent combination of the quantum postulate with classical mechanics required that some piece of the classical framework would have to be given up in order to obtain a harmonious framework for describing atomic phenomena. The light-quantum idea seemed to him to embody this inconsistency in the fact that it used the field representation of light to define a particle representation of the same thing. Thus he was inclined to reject this way of explaining the interaction between matter and radiation, but that implied that energy conservation would have to go. As Klein recalls,

> ... [Bohr] didn't want to think that light quanta were as real as atoms, and the backing for that was of course that they couldn't be defined without the waves. Therefore he wanted to keep the wave theory as something rigorous for light. ... Another thing must of course enter in—that energy conservation didn't belong to those physical laws of which he had the very definite opinion that they were quite fundamental. ... There were certain things that he thought must be quite fundamental ... but apparently the energy principle for him did not belong to these necessary foundations. I think that it was not that he wished to give up the energy principle, but the quantum paradox pressed on all of us to give up something.[21]

Thus we find that early in 1924 he wrote A.A. Michelson in a rather optimistic frame of mind that

> ... it appears possible for a believer in the *essential reality* of the quantum theory to take a view which may harmonize with the *essential reality* of the wave conception ... [and] it seems possible to connect the discontinuous processes occurring in the atoms with the continuous character of the radiation field in a somewhat more adequate way than hitherto perceived.[22]

Here Bohr is searching for a way to harmonize the quantum description of atomic systems with a belief in "the essential reality" of the waves-in-a-field representation of radiation. But of course the price to pay for this harmony would involve abandoning strict energy conservation in individual interactions. Most physicists would not have been willing to pay the price, but apparently Bohr was an exception.

If abandoning energy conservation was the answer, then the quantum description of the atomic system as changing its state discontinuously could be understood as a true representation of "the essential reality" so described. Bohr notes this possibility in a letter to Fowler written in late 1924:

> After all I believe that there may be more truth in the pseudo-mechanical treatment I tried in old times than one might perhaps think. In fact I believe we have here to do with an instructive example of the limitations in the ordinary quantum theory rules ... which affords an illustration of the necessity of giving up the strict validity of the general principles of conservation of energy and momentum.[23]

Here Bohr indicates that faced with the choice between giving up understanding the quantum description of atomic systems and the field description of radiation as representations of "the essential reality" or giving up strict energy conservation, Bohr's natural tendency was to sacrifice the latter and stand by an interpretation of the theory which treats its representations as descriptions of real entities rather than heuristic constructions.

However, five months after this letter was written to Fowler, an experiment by Walter Bothe and Hans Geiger in Germany demonstrated that energy was strictly conserved in individual atomic interactions, thus disproving Bohr's proposal. The very day Bohr received news of this experiment, he wrote Fowler again,

> ... it seems therefore that there is nothing else to do than give our revolutionary efforts as honorable a funeral as possible. ... In fact I think that the possibility of describing these experiments without a radical departure from an ordinary space–time description is so remote that we may as well surrender at once and prepare ourselves for a coupling [*i.e.*, an interaction] between the changes of state in distant atoms of the kind involved in the light quantum theory. ... I am thinking of all kinds of wild symbolic analogies.[24]

This letter gives testimony to two important ideas present in Bohr's thought as he recognized that harmony was not to be achieved by abandoning strict energy conservation. First he realized that since energy was conserved, the description of the interaction between matter and radiation could not represent radiation *in these phenomena* as a continuously changing field. Thus he is suddenly prepared to take the light-quantum theory much more seriously than before. He now begins to believe that with respect to radiation the field representation was not going to be victorious over the merely "heuristic hypothesis" of light-quanta. Second, he realized that since energy was strictly conserved, the classical application of the claim of causality through using the conservation principles would have to be retained.

Thus the other aspect of the classical ideal for a description of a physical system, that of space–time description, now becomes suspect. Consequently it is at this point that Bohr begins to consider the possibility that the theoretical representations of isolated atomic systems through the spatio-temporal "picture" of a system of particles moving on definable trajectories and of radiation in free space as a wave moving through a continuous electromagnetic field cannot be understood as an "ordinary space–time description" (*i.e.*, in the classical sense) of the physical systems to be described. Once he realized that energy conservation could not be given up, the dualism of particle and wave representations of electromagnetic radiation became a fact with which the new framework would have to reckon. For this reason, he now concentrated his attention on how such "wave" and "particle" space–time descriptions were related to the systems they were used to represent.

4. Discussions with Heisenberg and Matrix Mechanics

During the spring of 1925, while Bohr was burying his revolutionary efforts at abandoning energy conservation, he was engaged in daily discussions with his brilliant young assistant Werner Heisenberg, then working at Bohr's Institute. Heisenberg recalls that these discussions influenced his formulation of "matrix mechanics" later that June, but they had an impact on Bohr's thought as well.

We have just seen that the failure to discard energy conservation now focused Bohr's interests on an interpretation of "space–time descriptions". The notion of defining the state of atomic systems such that one could describe a "picture" or "model" (Danish "*billede*" or German "*Bild*") which represents the spatial loci of the component particles at each temporal instant was of course an extension of the classical ideal of description into the atomic domain of which Bohr had made use in his 1913 atomic model. We know from his manuscripts and letters of this period that Bohr continually puzzled over how a space–time description of the atomic system, which was defined in terms of parameters which are continuous, could be used to define the state of a system which changed discontinuously. In the second of the letters to Fowler quoted in the previous section, his new-found suspicion concerning space–time description stands in marked contrast to his hopes in 1924 (when non-conservation of energy seemed a real possibility) of taking more seriously the "truth in the pseudo-mechanical treatment" he had tried in his original atomic model.

Hence it comes as no surprise that Heisenberg recalls that in their discussions during the spring of 1925, Bohr reluctantly agreed for the first time to completely abandon any attempt to describe the atomic system in terms of "visualizable" or pseudo-mechanical models.[25] Here "visualizable" clearly refers to "space–time description" as classically understood. Nevertheless, it is probable that the parties to this agreement had rather different interpretations of what had been agreed to. On the one hand Heisenberg apparently read their agreement as Bohr's endorsement for pursuing a purely mathematical theory that would ascribe properties only to observed phenomena resulting from interactions between atomic systems and radiation.

On the other hand, Bohr characteristically read this same agreement as endorsing a search for a revised understanding of how we use space and time concepts in picturing the behavior of atomic systems.[26] For Heisenberg this resolve helped to produce first matrix mechanics, then some twenty months later, the uncertainty principle. For Bohr this agreement marked a major step on the road to complementarity.

The record of this agreement is of some interest, for it reveals that following the collapse of the attempt to discard strict energy conservation, Bohr's analysis of space–time description proceeded quite independently of Heisenberg's train of thought leading to matrix mechanics and the uncertainty principle. Heisenberg saw Bohr's doubts about "visualizable models" as implying that a theoretical representation of the atomic system could and should proceed without attempting a space–time description of the physical conditions actually obtaining within the atomic system. Heisenberg concluded that the theory should instead focus on simply predicting the results of interactions between radiation and the atomic systems. He recalls that, "This developed into a long and heated discussion during which ... the necessity for detachment from intuitive [*i.e.*, "visualizable" in the sense of a space–time description] models was for the first time stated emphatically and declared to be the guiding principle for all future work."[27] According to Heisenberg's recollection, in these discussions Bohr and Kramers first agreed with Heisenberg's approach, in which a magnetic field is fictionalized for the sake of arriving at the right formula to predict observed phenomena, and then this device of the magnetic field is admitted to be purely fictional and discarded. Since Heisenberg believed that only a mathematical formalism for predicting observed results was needed, he saw nothing objectionable in this approach. But after first agreeing, Bohr grew critical of Heisenberg's disregard for describing the physical situation in the atomic system. "I was completely shocked", recalls Heisenberg; "I got quite furious because I thought I had something real, and now they tried to explain it away. So we had quite a heated discussion, but at the end I think I came out with a slight victory. ... And I had for the first time the feeling that now I had been able to convince Bohr about something about which we had disagreed."[28]

Early that summer Heisenberg retired to the island of Heligoland to recover from an attack of hay fever. During this time he succeeded in formulating what his Göttingen mentor, Max Born, later realized was a matrix calculus making possible predictions of the relevant observables in the interactions between atomic systems and radiation. This matrix mechanics was the first expression of the "new" quantum mechanics, a theory which completely eradicated any dependence on a classical space–time description of the atomic system, but provided a logically consistent mathematical formalism for predicting the results observable in interactions between atomic systems and radiation. Heisenberg represented the probabilities of atomic systems transferring from one stationary state to another without reference to any space–time picture of the atomic states in terms of electron orbits or other such pseudo-mechanical representations. But the conceptual link needed to connect this purely formal mathematical scheme with the description of physical processes involved in such interactions remained as obscure as ever.

By intensifying the inadequacy of the conceptual framework of classical mechanics, Bohr had been led to his original atomic model. So in this case the success of matrix mechanics stimulated him to determine the exact point where the classical descriptive ideal broke down. This search led ultimately to complementarity; so in both cases Bohr's great insights resulted from a similar method of discovery.

Bohr's concern to represent in his imagination a model of the physical situation which in the observational interaction causes the phenomena predicted by the mathematical formalism did not mean that he regarded such a representation as literally a picture of the atomic system. But he wanted to know how the classical concepts used to describe the experimental phenomena related to the atomic system which was theoretically represented as producing such phenomena. Bohr understood more profoundly than Heisenberg that the scientific description of nature requires more than just a mathematical formalism for prediction of observed phenomena. In order to understand how to apply the mathematical formalism to give predictions of what will be observed, it is necessary to be able to represent conceptually what happens in the physical interactions which are described as "observations" determining the properties of atomic systems. Undoubtedly Bohr had expected that a new framework providing just such a conceptual representation for describing the atomic system would have to appear *before* physics advanced to a mathematical formalism capable of predicting what would be observed. However, Heisenberg's discovery reversed the historical order Bohr had anticipated. Heisenberg admits that "it was probably a surprise for him [Bohr] that there could be just one new mathematical scheme which would do the whole thing in one step".[29] But Bohr's surprise at Heisenberg's formalism was justified, for he knew what Heisenberg had yet to learn: for a phenomenon to be regarded as an "observation" it is necessary to describe a physical interaction; it does not count as an observation *of* anything until we have some means for describing the physical situation as an observation of some object. What we use to describe such observations are in fact the concepts of classical physics. But an interpretation of these *classical* concepts as referring to properties of the objects described leads to paradoxes such as the dualism of wave and particle representations of radiation. To remove such inconsistencies a new framework within which we understand this use of descriptive concepts must be formulated. Such a new framework must alter our understanding of how these descriptive concepts relate to the atomic systems the behavior of which they are used to describe.

Consequently, although Heisenberg was understandably overjoyed by his great discovery, Bohr was far from content, as he should have been had he embraced the view that all that was needed was a predictive scheme. His discontent at this time belies the belief that the quantum revolution implies abandoning the attempt to understand the properties of atomic systems as real entities and the resulting misunderstanding of complementarity. Bohr's distress in 1925–26 demonstrates the creative tension which encouraged him towards a renewed search for a conceptual framework within which Heisenberg's formalism could be understood. Heisenberg recalls this tension:

> Bohr, I would say, was a bit uneasy about the whole thing. He felt very strongly, "Well there must be something in it", but at the same time he also saw that one had not solved the problem of using the right words [*i.e.*, concepts]. This way of attack—that first you must have a mathematical scheme, then you will find the language—was ... just the opposite of his own attitude. Therefore he was very much worried and then hoped that one should learn from the mathematical scheme what it was all about. He was not so much interested in a special mathematical scheme.[30]

However, the highly abstract character of matrix mechanics seemed to bar the way to finding any physical interpretation of the mathematical scheme. Thus Heisenberg's achievement led Bohr to analyze the relationship between the empirical reference of the classical descriptive concepts and the meaning of those same concepts within the quantum theoretical representation of atomic systems.

The classical concept of "spatial position" is a good example to illustrate what puzzled Bohr. It can be given an empirical reference through specific operational procedures which locate the position of a particle in a three-dimensional empirical space. In classical mechanics the same concept is given theoretical meaning by representing position as a point in a three-dimensional theoretical space. Thus if the theoretical representation is interpreted as describing the properties of a real entity, no inconsistency arises in regarding the theoretical parameter of spatial position as corresponding to the property of the particle located at a point in a real three-dimensional space. Furthermore, this property of the particle may be regarded as the mechanical cause of the observational interaction in which a measuring system "observes" the particle as at this point.

However, in Heisenberg's new mechanics the concept of position retains its customary classical meaning with respect to observation, but as a theoretical parameter it is represented as a matrix array of many numbers. Thus there arises a puzzling inconsistency between the theoretical representation and the classically defined empirical reference of "observed" position in terms of a single determinate point in three-dimensional space. This inconsistency obscures how such a theoretical parameter can "correspond" to the property of the system so described. It was precisely the desire to understand the properties of real atomic systems which made this obscurity so frustrating to Bohr. Heisenberg gropes to explain his and Bohr's state of mind at the time:

> When you speak about the model, you mean something which can only be described by means of classical physics. As soon as you go away from classical physics, then, in a strict sense you don't even know what a model could possibly mean because then the words haven't got any meaning anymore. Now this was a dilemma. ... Bohr tried to keep the picture while at the same time omitting classical mechanics. He tried to keep the words and the pictures without keeping the meanings of the words of the pictures. Both things are possible in such a situation because your words don't really tackle the things anymore. You can't get hold of the things by means of your words, so what shall you do? ... Bohr's escape would be into the philosophy of things.[31]

This "escape" into philosophy meant analyzing the way in which the concepts apply to describing nature, and that in turn meant taking seriously the dualism of wave and particle representations of light.

Certainly, since Bohr had already anticipated the possibility that the new mechanics would have to surrender space–time "pictures", matrix mechanics must have confirmed his suspicions. But without any physical interpretation of the mathematical parameters of this representation, we could not say that we had "described" the atomic system, at least certainly not in accord with the ideal for a description in the classical framework. What was needed, then, if the theory was right, was a new ideal for an adequate description, and this in turn meant a new framework. But how could such a descriptive ideal be expressed if the classical concepts which must be used to communicate the results of observation no longer have their old meanings? This was the problem as Bohr now came to see it. Following the spirit of his correspondence principle, his guide was to generalize the classical framework so that the new framework with its ideal would preserve those successful descriptions of classical physics and yet allow for a new description of atomic processes through matrix mechanics as well.

5. Wave–Particle Dualism

By the mid-twenties the pace of theoretical discoveries had accelerated rapidly. Most importantly, Louis de Broglie made the theoretical suggestion that systems whose behavior could be described previously only by representing them as composed of particles could also exhibit phenomena that could be described only by representing these same systems as like waves in a field. Just as Einstein had used Planck's quantum idea to suggest a particulate representation of electromagnetic radiation which could predict some phenomena that the field representation could not, so de Broglie turned the same idea around and suggested that by representing what were previously thought to be particles as wave-like, theory could be used to predict hitherto unobserved wave phenomena for what had been thought to be particles. Specifically this new representation predicted that certain interactions involving electrons, which were previously thought to *be* particles, would produce phenomena that could be described only by representing electrons as a wave disturbance in a field. These predictions were in fact confirmed two years later by C.J. Davisson and L.H. Germer.

Thus by 1925 when matrix mechanics appeared, both wave *and* particle representations seemed necessary to describe the full range of phenomena in which atomic systems and radiation were said to be observed. However, if one understood the parameters of these theoretical representations as corresponding to the properties of real objects thus described, then it would seem that these systems must have contradictory properties. Bohr distilled this inconsistency into the "dualism" of particles and waves.

The thrust of his efforts, therefore, lay in exposing the disparity between the fact that on the one hand in order to describe atomic systems there already existed a coherent mathematical scheme for making predictions confirmed by observation. But on the other hand the physical interpretation of this formalism, viewed classically, seemed to require that "nature commits a contradiction" by having an entity with the logically incompatible properties of being at a determinate point in space (a particle) and at the same time being spread through a region of space (a wave-in-a-field). In line with his method of intensifying the conceptual breakdown, Bohr tried to design a "thought experiment" which would "catch nature in a contradiction". Heisenberg recalls, "Bohr would ask, 'Well how can nature avoid contradictions? Now we know the wave picture, we know interference, ... we know all that—how can our Lord possibly keep this world in order?' "[32]

Of course this paradoxical dualism had puzzled physicists ever since Einstein's photon hypothesis of 1905, but the very fact that at such an early date atomic phenomena were so little understood, made the answer to the dilemma seem so remote that it was possible to push it into the background. Thus physicists had grown use to the inconsistency of moving back and forth from classical to quantum ideas. Heisenberg describes their outlook as follows:

> When the physicists had met the existence of the quantum ... people were able to feel that this was something important and very new. But they were not able to do the other step which would have been absolutely necessary to come further, and that is to throw away the old physics. ... Then what could the physicists do? They would of course try to use the old concepts and try to add if possible these new ideas in places where they found them necessary. ... Bohr, for instance, in his model of the atom, just used this idea of discontinuity to explain one essential fact, namely the stability of the atom. It took twelve years until one dared to really go away and push all the old concepts aside. ... You must remember that not only the tradition of two hundred years of physics, but also all the experience of the planetary motions and everything else just proved that classical mechanics is right.[33]

For twelve years physicists could live in a world ruled by this monstrous dualism only because the dualism was so monstrous and offered so little hope of solution that people learned to live with it. Of course this contradiction could be easily solved by rejecting the interpretation of the goal of theory as the description of the properties of real atomic systems. But such an easy exit would mean abandoning any hope of describing how the interaction of the observing instruments with the atomic system produce the observed phenomena which confirm the mathematical scheme. And this was a price that Bohr, for one, was not prepared to pay.

The younger physicists who had not known a time before the dualism appeared "had come into the habit of playing with different pictures. ... [they]

asked 'What would nature do if we asked her this question?' ... everybody hoped that someday somebody would find the real picture which is behind it. ... the playing between different pictures was meant as a study to prepare this new thing to happen."[34] But before 1925 Bohr was still very reluctant to accord equal status to the light-quantum idea and de Broglie's electron-wave idea had not yet appeared. Thus he chastened the younger men, forcing them to come to grips with the paradoxes to warn them that they could not so light-heartedly accept the reality of the light-quantum and still use the wave theory for electromagnetic radiation. Once he saw that the light-quantum concept was with physics to stay, he emphasized the paradoxes even more. Thus as more and more of the quantum jigsaw puzzle was completed, Bohr became ever more aware of the absence of the most crucial piece: the framework necessary for relating the descriptive concepts to the objects which he regarded the theory as intending to describe. Heisenberg rememebers Bohr's methodical preoccupation with "those hopeless paradoxes":

> Those paradoxes were so in the center of his mind that he just couldn't imagine that anybody could find an answer to the paradoxes, even having the nicest mathematical scheme in the world. ... The very strange situation was that now by coming nearer and nearer to the solution the paradoxes became worse and worse. That was the main experience. ... nobody could know an answer to the question, "Is an electron now a wave or is it a particle, and how does it behave if I do this or that and so on." Therefore the paradoxes became so much more pronounced in that time. ... only by coming nearer and nearer to the real thing to see that the paradoxes by no means disappeared, but on the contrary got worse and worse because they turn out more clearly. ... like a chemist who tries to concentrate his poison more and more from some kind of solution, we tried to concentrate the poison of the paradox. ... [35]

Bohr's unique sensitivity to the need for using classical concepts to describe the physical situation in an observational interaction and his clear awareness of the revolutionary consequences of the quantum postulate prevented him for settling for a purely abstract formalism and made him acutely aware that no "cheap solutions" were to be had by returning to the classical ideal of continuous change of state. However, they also gave him a seriously divided outlook toward just how fruitful this inconsistent combination of ideas would turn out to be. Prior to his final acceptance of the light quantum hypothesis on the basis of the Bothe–Geiger experiment, he reveals this attitude in a letter to Pauli, always a sharp critic of logically muddled thinking:

> Perhaps I ought also to have a bad conscience with respect to the radiation problems, but even if from a logical point of view it is perhaps a crime, I must confess that all the same I am convinced that the swindle with mixing classical theory and quantum theory will still in many ways show itself fruitful for tracking nature's mysteries.[36]

We have seen that by 1925 Bohr came to believe that the key to the new framework which would resolve the inconsistencies between the classical and quantum ideas was to work with *both* particle and wave pictures until one could understand how these concepts related to quantum systems. Heisenberg remembers being alienated by this way of thinking because he saw the use of these classical "pictures" as clinging to the old physics which was destined to be replaced. For him the consistency of the mathematical scheme guaranteed that the new line of approach was not involved in any contradictions:

> Bohr was very much inclined to go forth and back between wave and particle pictures. That was a thing which I didn't like too much because I felt that at least quantum theory [matrix mechanics] seems to be a consistent scheme and so we should be able to talk about nature only by using this scheme and not introduce other schemes. ... for me it was clear that ultimately there was no dualism. ... [37]

Bohr would have agreed that in the final analysis there could be no dualism, but he saw that the the classical concepts of particle and wave were necessary to describe the physical interactions which produce the observed phenomena that confirmed the mathematical formalism, and by 1925 he was convinced that neither was to be discarded in favor of the other. Thus what was needed was not a victory for the particle picture or the wave picture, but rather an explanation of how the apparent contradiction involved in this dualism resulted from a misunderstanding of how these concepts relate to the systems they are used to represent.

In December of 1925 Bohr's commitment to dualism increased as a result of discussions with Einstein in Leiden on photon versus wave representations of radiation. By this time Bohr was certainly taking the photon concept much more seriously than a year earlier. For his part, Einstein seemed to recognize the indispensability of the wave representation.[38] Earlier that July Bohr had cautioned in an addendum to another article that though the Bothe–Geiger experiment had disconfirmed his suggestion that energy was not strictly conserved in atomic interactions, it did not establish unequivocally the reality of the photon. Instead, he argued that this experiment demonstrated the need for further revolution: "One must be prepared to find that the generalization of classical electrodynamic theory that we are striving after will require a swelling revolution in the concepts on which the description of nature has been based up to now."[39] By that October, his disagreement with Einstein over the photon was so well known that it reached the newspapers, although the two men had not yet met in Leiden or even exchanged any correspondence on this issue.

However, after the Leiden meeting Bohr began to indicate that he had finally resigned himself to accepting the unavoidability of the light-quantum representation, though of course he pointed out that Einstein must also accept the unavoidability of the wave representation as well. In early 1926 he wrote to John Slater:

> In Leiden I have recently had a long discussion with Einstein about the state of radiation theories. Although we were wrong [about energy non-conservation] ... in which respect I have a bad conscience about persuading you to our view, I believe that Einstein agrees with us in the general ideas, and that especially he has given up any hope of proving the correctness of the light quantum theory by establishing contradictions with the wave theory description of optical phenomena. In my opinion the possibility of obtaining a space time picture based on our usual concepts becomes ever more hopeless.[40]

Note that here again Bohr's acceptance of wave–particle dualism increased his suspicion concerning the "possibility of obtaining a space time picture based on our usual concepts". Three days later he expressed the same suspicion concerning mechanical models in a letter to Oseen, where he mused, "in every result the temptations to go astray are waiting", meaning that it may have been *bad* fortune that his early model was so successful in combining classical mechanics with the quantum postulate, for "had the connection only been limited ... [we would not have] been tempted to so crude an application of mechanics as was for some time thought possible".[41] Thus 1926 seems to mark the lowest point in Bohr's very real despair over ever successfully describing the atomic system.

6. Discussions with Schrödinger and Wave Mechanics

In most quarters the success of Heisenberg's matrix mechanics was hailed as the final attainment of a consistent description of the atomic system. Thus all were stunned when only a few months after Heisenberg's theory, a totally different approach yielded an equally consistent means for predicting the same phenomena. This was the contribution of Erwin Schrödinger.

Schrödinger's wave mechanics, as his new theory was called, was an important departure from the previous direction of progress. Following the discovery of de Broglie that the electron could be represented as a wave, Schrödinger sought an equation representing the state of the atomic system as a wave in a field, or more accurately, as the superposition of a series of waves. Schrödinger regarded his wave equation as eliminating the quantum problem by reestablishing a classical representation of the system as changing its state continuously. However, it was soon realized that the wave of Schrödinger's equation could not be interpreted as representing a real wave in the three dimensions of physical space, for the wave equation required three distinct dimensions for each particle of the system so described. Thus Schrödinger's new mathematical scheme revealed no more of the physical processes occurring when an atom transfers from one stationary state to another than had Heisenberg's matrix mechanics. In fact, soon the two theories were shown to be mathematically equivalent.

In September 1926 Schrödinger visited Bohr's Institute for a round of intense discussions on the physical interpretation of the new theories. The presence of two mathematical formalisms, one developed from a particle approach and the other from a wave approach, had intensified the wave–particle paradox. Thus the discussions with Schrödinger proved a vital forum for clarifying Bohr's quandary. Recalling that Bohr was generally incapable of developing his ideas alone, we can well appreciate that the presence of Schrödinger as a sounding board provided a necessary ingredient for Bohr's creative dialectic. Heisenberg remembers the intensity of the discussions:

> It was almost a pity to see; I mean Schrödinger had given this talk and he was criticized by Bohr. ... In some way Schrödinger was so worried that he got ill there. ... Bohr would sit at Schrödinger's bed and would say, "Now Schrödinger you *must* see, you *must* see", you know. I mean he would not let him go for a single minute in the whole day; he wanted all this push, push, push, you know. So that's Bohr, Bohr wanted complete clarity to the end. ... Bohr was terribly anxious to get to the bottom of things. ... I think he [Schrödinger] was really in a kind of despair, he said in the end that if one had to stick to those darn quantum jumps then he regrets that he had ever taken part in the whole thing. In some way he was extremely angry about this outcome, but he could not defend his position. He saw, at least at that time, that he could not really defend a new explanation. ... [42]

In the development of complementarity Bohr's discussions with Schrödinger, and his subsequent analysis of these discussions with Heisenberg from September 1926 to February 1927, marked the final tightening of the creative tension between quantum discontinuities and classical continuity that led Bohr to his second great insight. The impact of Schrödinger's visit is obvious in several of Bohr's letters. To Fowler he wrote: "I think that we have succeeded in convincing him [Schrödinger] that for the fulfillment of his hopes [*i.e.*, to avoid those 'darn jumps'] he must be prepared to pay a cost as regards reformation of fundamental concepts, formidable in comparison with that hitherto contemplated by supporters of the idea of a continuity theory of atomic phenomena."[43] One month later in a letter to Darwin, Bohr explicitly links Schrödinger's visit to his ever more sure conviction that we must deal with both continuity (a field description) and discontinuity (his particle model of atomic systems):

> ... about a month ago we had a very interesting visit of Schrödinger which gave rise to quite heated discussions, none of us here being quite prepared to follow him in his hopes to establish a proper continuity theory. Of course the results Schrödinger has got are so marvelous that we all appreciate his hopeful state of mind in this respect, but at the same time I think that he had underrated the fundamental nature of the difficulties we are up against. ... it is very interesting to see how the action of a corpuscle or wave presents itself as the more convenient according to the place in the construction where the departure of discontinuity involved in the postulates is introduced. Indeed this is only what might be expected, since every notion, or rather every word, we use is based on the idea of continuity, and becomes ambiguous as soon as this idea fails.[44]

The last sentence is particularly interesting, for it expresses a point Bohr was to make time and time again using almost precisely the same wording: the classical concepts presuppose continuity in change of state for their use in describing interactions between physical systems. When that presupposition is denied, the use of these concepts becomes "ambiguous".

Although Bohr reacted to Schrödinger's contribution with increased distress over the quantum paradoxes, Heisenberg took the mathematical equivalence between his and Schrödinger's formulations as further confirming his optimistic attitude that the situation was nearly fully understood. Consequently he had to be reminded by the ever scrupulous Bohr that the "yawning abyss" remained between the classical conception of systems changing state continuously and the quantum description of them as changing state discontinuously. As long as this fact and that of wave–particle dualism remained so ill understood, the relation between the theoretical formalism and the description of atomic processes was not understood. Heisenberg remembers with a touch of irony that

> ... sometimes Bohr and I would disagree because I would say, "Well, I'm convinced that this is the solution already." Bohr would say , "No, there you come into a contradiction." Then sometimes I had the impression that Bohr really tried to lead me onto "Glatteis", onto slippery ground, in order to prove that I had not the solution. But, this was, of course, exactly what he had to do from his point of view. It was perfectly correct. He was also perfectly correct in saying, "So long as it is possible that you get onto slippery ground, then it means that we have not understood the theory." I remember that I was sometimes a bit angry about it, which was, of course, natural.[45]

The difference between the intellectual demands of both physicists led to their different creative efforts to resolve the problems. Heisenberg was content that the consistency of the formalism assured that an understanding of atomic systems had been achieved. Consequently he believed that all that was left was to clarify the application of the formalism to concrete situations. Of course Bohr demanded much more. Heisenberg's attempts to escape Bohr's challenges led him to discover the uncertainty principle as a way of showing that nature cannot be caught in a contradiction. Bohr's overarching desire for a conceptual framework stipulating a new ideal for an acceptable description of atomic phenomena led him to complementarity. The following two sections will consider each intellectual path in turn.

7. The Discovery of the Uncertainty Principle

In trying to answer how the mathematical consistency of matrix mechanics could guarantee that it was always possible to avoid Bohr's challenges, Heisenberg had the inspiration to "turn the question around", asking not how does the mathematical scheme copy nature, but rather why it is that nature allows only those physical situations which can be represented in the mathematical scheme.

The summer of 1926 afforded Heisenberg the opportunity to meet Einstein and put before the elder physicist the way in which he thought his new matrix mechanics embodies the same sort of thinking that Einstein had used to present relativity:

> I told him [Einstein] that this idea of observable quantities was actually taken from his relativity. Then he said, "That may be so, but it is still the wrong principle in philosophy." And he explained that it is the theory finally which decides what can be observed and what can not, and therefore, one cannot, before the theory know what is observable and what is not. ... On the other hand, of course, he agreed that as a heuristic principle it was extremely important. It was a way of finding out what one should put into a theory.[46]

Heisenberg appears to have taken to heart Einstein's advice, for after a heated debate with Bohr some time in early January of 1927, in a "half-angry" state of mind over Bohr's latest challenge, it occurred to Heisenberg to ask, "why not simply say that only those things occur in nature which fit with our mathematical scheme".[47] By early February both men were in a "kind of despair" and agreed that each needed to think alone; so Bohr went on a skiing trip to Norway while Heisenberg remained in Copenhagen. Bohr returned from that trip with the idea of complementarity, while Heisenberg succeeded in distilling the uncertainty relations from their discussions:

> I tried to say what space meant and what velocity meant and so on. I just tried to turn the question around according to the example of Einstein. You know Einstein just reversed the question by saying, "We do not ask how we can describe nature by mathematical schemes, but we say that nature always works so that mathematical schemes can be fitted to it." ... Therefore, I just suggested for myself, "Well, is it not so that I can only find in nature situations which can be described by quantum mechanics?" Then I asked, "Well what are these situations which you can define?" Then I found very soon that these are situations in which there was this Uncertainty Relation.[48]

In order to explain why Heisenberg's discovery represented a shock to the classical view, it will be convenient to pause briefly to introduce the formulation of classical mechanics developed by W.R. Hamilton in the early nineteenth century. According to Hamilton's method, the state of an isolated system may be represented by means of a function (the "Hamiltonian") of two parameters from which one can determine the future state of that system for any instant as long as it remains isolated. The specific form of the Hamiltonian function depends on the nature of the system, but in each case it is a function of a pair of state parameters one of which is used in the mode of space–time co-ordination and the other of which is necessary for the application of the conservation principles. These pairs of parameters are said to be "canonically conjugate". There are several such pairs, but to explain the uncertainty relations, it is most useful to concentrate on the parameters which define the state of a particle moving in space through time, those of position and momentum. The parameter corresponding to the position of the particle may be determined by an observation. But in order for the information thus obtained to be used to define the state of the isolated system, it is necessary to use the conservation principles to determine the effect of any "disturbance" caused by the particle's interaction with the instruments used for observing its position. Clearly in the case of visually observing large objects, the disturbance in the object's state caused by the observation is in fact negligible. But even in cases where it is not negligible, the classical framework permitted compensating for the effect of such a disturbance because the way in which the state was disturbed could be determined by mechanical theory. Bohr expressed this fact by saying that in the classical framework interaction was "controllable".

To define completely the classical state of the system, it is necessary also to determine by a different observational interaction the momentum parameter for the system at the same instant as that for which the position has been determined. But such a momentum determination requires interacting with the observed system in a way which physically precludes the conditions necessary for an observational interaction which would determine the particle's position. Thus even in the classical framework both canonically conjugate parameters cannot literally be simultaneously determined by *observation*. However, this fact does not make unattainable the classical descriptive ideal of *defining* the mechanical state of the particle, because we can make position and momentum observations successively in time. Then we can use classical mechanical theory to determine the value of the first parameter at the same time that the second parameter was determined. Thus in the classical framework one can consistently speak of defining the system's state at the moment it is observed.

Of course momentum is not "observed" in the same sense as position is observed. What is literally observed in a momentum determination is the spatial position of some piece of measuring apparatus which is so designed that such a space–time description of it reveals how its state was changed by the momentum of the particle with which it interacted. In this way, using observation and theory, the classical framework can accept defining the state of an isolated system through two canonically conjugate parameters as an attainable goal or ideal of description. But it was able to do so only by presupposing that the observational interactions used to determine the system's state changed that state in a way which could be completely determined by mechanical theory. But using classical mechanical theory to determine the change produced by such an interaction required that the properties corresponding to the canonically conjugate parameters changed *continuously* through time. For this reason, in order to make the classical descriptive ideal of defining the mechanical state of the system as isolated from interaction consistently attainable by observing the system, it must be presupposed that the mechanical state of the system changes continuously.

The "uncertainty" relations express the mathematical consequence of the quantum mechanical formalism that it is impossible to define the state of a physical system by precise values of the canonically conjugate parameters which define its classical mechanical state. Because of this fact, it is not possible within the quantum mechanical representation to describe a physical system by means of *both* a "space–time picture" giving the precise loci of all the system's components *and* a precise determination of the momentum and energy parameters necessary to apply the "claim of causality" through the conservation principles. One can make either description as precise as desired, but the more precisely one is determined, the greater the indeterminacy introduced into the canonically conjugate parameter. Thus in representing a system within the quantum mechanical formalism, the more precisely we realize what Bohr called the "ideal of observation" by determining the spatial positions of the system's components at an instant, the greater indeterminacy we introduce into the theoretical prediction of the future change of that system's state, or in other words the less precisely we are able to define the future state of the system by applying the claim of causality to achieve what Bohr called the "ideal of definition".

That this principle is a necessary consequence of representing systems in the way they are represented in the quantum formalism is not controversial. However grave differences of interpretation arise over why this limitation is imposed on the description. Heisenberg tried to show that it expresses the physical fact that, given the discontinuous change of state implied by the quantum postulate, it is impossible to arrange a physical situation permitting an observational interaction which simultaneously determines the precise values of those phenomenal observables which provide the empirical reference of both canonically conjugate parameters. To show this he devised the famous thought-experiment with a "gamma-ray microscope" through which one observes the position of a particle. The effect of the quantum representation of the interaction between the particle and the microscope used to determine its position is such that the more precisely position is determined the less precisely one can determine the momentum.

A consequence of this mode of presentation was to suggest the possibility that the particle does in fact have, or exist in, a well-defined classical state, *i.e.*, it possesses both position and momentum, but that due to the observational interaction it is impossible to *know* these parameters with equal certainty. Consequently, those dissatisfied with the quantum mechanical representation tried to attack the theory by devising a way in which both parameters could be determined, at least in principle. As we shall see, in order to defend against such attacks and argue that the quantum mechanical representation was "complete" (*i.e.* that a more precise definition of state was impossible) Bohr was forced to a much more radical reading of Heisenberg's discovery.

Bohr came to understand the uncertainty principle when rightly understood as crucial support for his new framework. But Heisenberg's route in reaching it alarmed him, for Heisenberg had accepted the mathematical formalism first and then argued that nature always behaves such that it obeys it. The younger man knew that this approach would not be to Bohr's liking; thus, when Bohr returned from Norway, there followed two months of tense discussions. Heisenberg remembers that the first round ...

> ... ended with the general impression that now Bohr again has shown that my interpretation is not correct. Inside I was a bit furious about this discussion, and Bohr went away rather angry because he saw my reaction, whether I had expressed it or not. ... it ended with my breaking out into tears because I just couldn't stand this pressure from Bohr.[49]

By this time Bohr had worked himself towards the basic idea of complementarity by concentrating on the need to describe quantum systems through both particle and wave representations, but Heisenberg arrived at his conclusions by reasoning solely from his matrix formalism which had been derived purely from the particle description. Heisenberg took this to be the basis of Bohr's suspicions concerning his approach:

> [Bohr] had thought himself into this scheme of dualism. He had come to the idea that we always needed two pictures. ... He would simply say, "Well have I *understood* how the thing works?" And he would say that to understand means to use the two pictures, waves and particles. ... Now here was a paper [*i.e.* Heisenberg's draft of the uncertainty relations paper] which tried to go entirely on the one line, namely the particle line ... so he felt, "Well this man forgets one half of the picture. If you only talk about the particle side you forget one half of it. It may be consistent but it's not the real story."[50]

Since Heisenberg's approach started from a purely formal mathematical representation of the quantum state of the system, at first Bohr failed to see any advance in understanding physically what is happening in atomic processes.

These disagreements arose because Bohr rejected Heisenberg's operational presupposition that nature must imitate a mathematical scheme. Bohr would have agreed with Einstein's admonition that "theory" determines what is observable, if "theory" is understood to include the conceptual framework within which the mathematical formalism is interpreted as representing the physical objects whose phenomenal behavior it is used to predict. But Heisenberg took Einstein's comment to refer to "theory" in the narrow sense of only the mathematical formalism. Since this abstract scheme was itself consistent, Heisenberg reasoned, within its description of phenomena, nature cannot be represented in contradictory ways. However, for Bohr such formal consistency was "trivial"; what mattered was the fact that viewed from the classical framework, quantum theory represented physical systems as having incompatible properties. What Bohr sought was a revised understanding of the relationship between the formalism and the use of classical concepts to describe the physical systems whose behavior it represents. It took Heisenberg some time to appreciate that unless one had done this, one had not, in Bohr's sense, *described* nature. Since he had discarded the classical framework and was content to work within a purely formal representation, he saw no need to concern himself with particle and wave "pictures", for these pictures were defined in terms drawn from the very framework which had been discarded. Heisenberg tells us that ...

> The main point was that Bohr wanted to take this dualism between waves and particles as the central point of the problem. ... I, in some way, would say, "Well we have a consistent mathematical scheme and this consistent mathematical scheme tells us everything which can be observed. Nothing is in nature which cannot be described by this scheme." ... Bohr would not like to say that nature imitates a mathematical scheme. ... While I would say, "Well, waves and corpuscles are, certainly, a way in which we talk and do come to these concepts from classical physics ... but since classical physics is not true there, why should we stick so much to these concepts? Why not say just that we cannot use these concepts with a high degree of precision, therefore the Uncertainty Relations, and therefore we have to abandon the classical concepts to a certain extent. When we get beyond this range of the classical theory we must realize that our words don't fit. They don't really get a hold in the physical reality and therefore a new mathematical scheme is just as good as anything because the new mathematical scheme then tells what may be there and what may not be there.[51]

It may have been the fact that Heisenberg had not experienced first hand the revolutionary break with classical mechanics which Bohr had struggled through in 1913 that caused him to perceive Bohr's concern with the classical concepts as a failure to relinquish one's hold on a now discarded means of description. However, Bohr's concern was not due to any lack of revolutionary zeal. His point was that in order to confirm the predictions of the mathematical formalism we must be able to represent the phenomena that provide the empirical evidence for the theory as observational interactions with physical systems in the atomic domain. And to do this, in fact we must employ those classical concepts which have unambiguous reference to the observable properties in terms of which such phenomena are described. Since those concepts cannot function as they had in the classical framework, until we can understand how they do function, we have not understood how the theory describes nature.

Since the acceptance or rejection of any mathematical scheme is at least partially a function of its successes or failures in predicting the outcome of observational interactions, it seemed obvious to Bohr that until one understood the role of the spatio-temporal observables which were used to describe such experimental outcomes one did not have a satisfactory account. Since in formulating such space–time descriptions we "picture" the physical system as a "wave" or a "particle", if we understand how the spatio-temporal concepts function to link theory to the phenomena, we should be able to say how such particle and wave descriptions apply to the systems which produce the phenomena. It was precisely because this is what Heisenberg could not say that Bohr was ever more pressed to search for a new framework, to "escape" in Heisenberg's words, "into the philosophy".

Bohr's initial concern with the uncertainty relations resulted from a fundamental error in Heisenberg's description of the gamma-ray microscope thought experiment. After Heisenberg's original analysis was corrected, a meeting of minds was achieved. But that meeting required that Heisenberg appreciate that the use of space–time concepts in representing particles and waves must be retained to describe the phenomena which the theory treats as the observations that confirm the formalism. Finally Heisenberg was forced to admit ruefully that

> ... just by these discussions with Bohr I learned that the thing which I in some way attempted could not be done. That is one cannot go entirely away from the old words because one has to talk about something. ... So I could realize that I could not avoid using these weak terms which we always have used for many years in order to describe

> *what I see*. So I saw that *in order to describe phenomena* one needs a language. ... Well we do have a language and that is the situation in which we are. ... we actually do use these precise terms and then we actually learn by quantum theory that we have used them in too precise a manner. The terms don't get hold of the phenomena, but still, to some extent, they do. I realized, in the process of these discussions with Bohr, how desperate the situation is. On the one hand we knew that our concepts don't work, and on the other hand we have nothing except the concepts with which we could talk about what we see. ... I think this tension you just have to take; you can't avoid it. That was perhaps the strongest experience of these months.[52]

By April of 1927 Bohr and Heisenberg finally reached an agreement, which Bohr marked by drafting a letter to Einstein to accompany a copy of Heisenberg's paper:

> Heisenberg has asked that I send you a copy of the proof he expects of a new article which he hopes will interest you. This article, which I am sending enclosed, probably marks a very momentous contribution to the discussion of the general problems of quantum theory. Since the contents are in close relation to the questions I have enjoyed discussing with you several times, most recently during the memorable time in Leiden at the Lorentz Festival, I would like to add a few comments in connection with the problems you discussed recently. ... It has long been recognized how intimately the difficulties of quantum theory are connected with the concepts, or rather the words, which are used in the description of nature and all of which have their origin in classical theory. Certainly these concepts give us only the choice between Scylla and Charibdis depending whether we direct our attention to the continuous or discontinuous side of the description. ... This situation permitted by the limitation of our concepts exactly co-incides with the limitations on our possibility of observations, in order to avoid all contradictions, as Heisenberg stresses. ... Through his new formulation we are given the possibility to harmonize the demand for conservation of energy with the wave theory of light, while in accord with the nature of description, the different sides of the problem never come into appearance simultaneously.[53]

This letter concisely includes many of the points which Bohr was to make in his arguments for complementarity, especially concerning the limitations in descriptive concepts resulting from their origin in the classical framework. It was to be a point he argued with Einstein many times, but he never successfully overcame the latter's skepticism.

8. Bohr's Second Great Insight: The Use of Concepts

As soon as agreement with Heisenberg had been reached, Bohr turned his full energies towards stating formally what he now believed to be the nucleus of a new framework for describing nature. The next chapter will consider that viewpoint; this section will analyze how his train of thought throughout this period took him to his position concerning the use of concepts, a view which in some ways was not new to Bohr at all, but was new to physics.

From the Bohr–Kramers–Slater paper to the first statement of complementarity was a period of three and one-half years during which Bohr published only two substantial papers. One of these, "Atomic Theory and Mechanics", may well be regarded as Bohr's first published philosophical effort to come to grips with the problems of the quantum mechanical description of nature. As we have seen, in recoiling from the failure of his suggestion of non-conservation of energy, Bohr began to ponder the classical goal of space–time description as a picture which locates the components of a physical system at determinate spatio-temporal positions at each instant of time. It is perhaps striking that in spite of the fact that matrix mechanics had just appeared, in this essay we find *no* attempt to deal with this mathematical representation of the atomic system, but find instead a strictly qualitative approach *about*, rather than *in*, atomic physics. This orientation indicates Bohr's expectation that a proper understanding of the atom would come only with understanding how the classical concepts functioned in a new framework.

Bohr began the paper by noting that in the Newtonian synthesis "the formation of concepts suitable for the analysis of mechanical phenomena was provisionally completed".[54] Then, he observes "the development of electromagnetic theories ... brought about a profound generalization of mechanical concepts. Although to begin with mechanical models played an essential part in Maxwell's electrodynamics, the advantages were soon realized of conversely deriving the mechanical concepts from the theory of the electromagnetic field."[55] Here Bohr is referring to his expectation that progress in describing nature can come from generalizing an older conceptual framework such that the successful descriptions achieved in it now appear in the new framework as a "special case" which is understood to occur only when certain special presuppositions not generally true are true in a particular instance. In this way his position concerning the description of nature was tied to his analysis of the way conceptual frameworks in science must change.

In what was to become a typical style of presentation, for most of the paper Bohr recounts a catalogue of achievements in atomic physics, which one by one spelled out the increasing inadequacy of what he here calls the "mechanical pictures". This expression clearly refers to "wave" and "particle" pictures in space–time descriptions. As we saw, Bohr's failure to describe successfully quantum interactions by abandoning the conservation principles had suggested to him that if we must keep strict energy conservation, we must abandon describing the interaction through space–time pictures. Given his whole frame of mind, it was inevitable that he approached this problem by asking how do the classical concepts, the "mechanical pictures", apply to the description of nature.

In the classical framework, if the theoretical representation of a physical system as a particle moving in a spatial trajectory is interpreted as picturing the properties of real entities, then the physical system really is a particle and possesses the properties that correspond to the parameters which define its theoretical representation. However, on this interpretation, wave–particle dualism seemed to imply that "nature commits a contradiction". Thus something must be wrong with such an interpretation, and consequently though these pictures cannot be discarded, for they provide the only way we have of describing phenomena as observational interactions with atomic systems, they can no longer be understood in the classical sense of providing visualizable models of how reality exists independently of our observation of it.

However, at this time Bohr was still unclear as to *how* the new framework would alter the classical use of these wave and particle pictures. He considers the possibility that wave and particle pictures might be abandoned altogether: "From these results it seems to follow that, in the general problems of quantum theory, one is faced not with a modification of the mechanical and electrodynamical theories describable in terms of the usual physical concepts, but with an essential failure of the pictures in space and time on which the description of natural phenomena has hitherto been based."[56] Nevertheless, Bohr does *not* accept this conclusion, for in the very next section of the paper he turns to an analysis of the way in which the correspondence principle—emphasizing as it does the continuity between classical and quantum predictions in descriptions where the effect of the quantum postulate is negligible—indeed does allow a *limited* application of space–time pictures: "Nevertheless, the visualization of the stationary states by mechanical pictures", he notes, "has brought to light a far reaching analogy between quantum theory and the mechanical theory."[57] Thus he seems tempted towards the conclusion, but does not *assert* it, that spatio-temporal pictures in the classical sense will not be abandoned *altogether*, but will be subjected to some *limitation* with regard to their applicability to the description of physical systems in the quantum theoretical representation.[58]

He continues this theme that the application of space–time concepts is limited with respect to atomic phenomena by mentioning only briefly Heisenberg's newly formed matrix mechanics,

> ... which has taken a step probably of fundamental importance by formulating the problems of quantum theory in a novel way by which the difficulties attached to the use of mechanical pictures may be avoided. ... In contrast to ordinary mechanics, the new quantum mechanics does not deal with a space–time description of the motion of atomic particles.[59]

But Bohr is apparently anxious *not* to discount totally the use of classical concepts by emphasizing that "the whole apparatus of the quantum mechanics can be regarded as a precise formulation of the tendencies embodied in the correspondence principle".[60] He then concludes his survey with the reaffirmation that "the fundamental difficulties involved in the construction of pictures of the interaction between atoms ... seems to require just that renunciation of mechanical models in space and time which is so characteristic a feature of the new quantum mechanics [*i.e.* Heisenberg's matrix mechanics]".[61]

Bohr's conviction that the new mechanics was on the right track seems to be linked to the fact that it *limited*, but did not *discard*, the classical concepts used to express the mechanical pictures. But the manner in which their use can be limited so that the descriptive ideals of both modes of description could be retained in spite of the quantum postulate remained obscure. At the time this article was written, 1925, Heisenberg had not yet discovered the uncertainty relations. So it had not yet occurred to Bohr that the two classical modes of description could both be retained if their descriptive roles were limited to describing the interactions between atomic systems which produce the *phenomena* that confirm the theory, rather than as referring to properties possessed by physical systems existing independently of observation. Nevertheless, his insistence on the value of the correspondence principle at this time does indicate that he rejected the temptation to *abandon* classical concepts in favor of some other non-classical concepts yet to be discovered. His basis for this attitude seems to be that without such concepts to describe what is actually observed in terms of space and time measurements, there is no connection between the theory and its empirical basis in observation. Furthermore, he stresses that were such concepts abandoned altogether, it would be hard to see how the classical wave and particle descriptions could be regarded as a special case of a wider conceptual framework which would "generalize" the classical framework in a manner suggested by the correspondence principle.

When Bohr combined his refusal to abandon the pictures altogether with his realization that one picture would not ultimately replace the other he was finally led to complementarity. The event which may well have precipitated this recognition was the appearance of Schrödinger's wave mechanics and the fact that on his visit to Copenhagen, Schrödinger was forced to admit (or so Bohr believed) that it was futile to continue attempts at a purely classical representation of the atomic system. Schrödinger's disappointment over the failure of his theory to reinstate classical continuity demonstrated his sensitivity to the crucial nature of this issue which other less perceptive thinkers ignored. But Schrödinger's intense desire to reinstate continuity stemmed from his wish to preserve the classical ideal of description. In contrast, Bohr's enthusiasm for revolutionary change was in reciprocal proportion to Schrödinger's disappointment and was equally a measure of his belief that the break in continuity physics required changing our understanding of the scientific description of nature in physics.

When Bohr returned from his Norwegian skiing trip in February of 1927 and was greeted with the news of Heisenberg's discovery of the uncertainty relations, it became clear to him that a new framework would have to retain *both* classical modes of description—that of space–time description with its wave and particle pictures and that of the claim of causality with its application of the conservation principles—but the two modes would have to be limited in their simultaneous application to the same physical system. Bohr saw the uncertainty relations as the theoretical expression of this limitation. But, as we have seen, he did not like Heisenberg's method of arriving at his discovery. Heisenberg's concentration on the consistency of the mathematical scheme led him to try to do without the mechanical pictures, and this is precisely what Bohr had concluded the new framework could *not* do. Since by this time Bohr did accept wave–particle dualism as unavoidable, he now concluded that these "pictures" must be used to describe the physical situation from which the theory permits predictions of what will be observed in specific observational interactions. But these wave and particle descriptions cannot be understood as picturing real entities possessing the corresponding properties apart from our observation of them. In considering countless thought experiments involving physical situations in which the observed phenomena could be represented as the consequence of an interaction in which the observed system must be represented as a *wave*, he and Heisenberg discovered that the physical conditions necessary to define this application of the wave picture always preclude those conditions necessary for an interaction which had to be described using the particle picture. Reciprocally, conditions necessary for the occurrence of those phenomena which could be described only by representing the system as a *particle* physically exclude conditions necessary for an observational interaction which would require representing the system as a wave. Since we can never observe a system as behaving in a way which forces us to characterize it as *both* a particle and a wave, it is perfectly unambiguous in the context of a theoretical description of a particular interaction to characterize the system as *either* a particle *or* a wave. No inconsistency arises from the dualism of particle and wave pictures as long as their reference is limited to describing particular *phenomena*. Thus "nature", in the sense of the field of possibly observable phenomena could not be caught in a contradiction.

For many scientists this situation was regarded as unsatisfactory because it was expected that new theoretical or experimental discoveries would reveal that the relevant phenomena could after all be described by a single representation, be it particle-like or wave-like in nature. For Bohr, too, this situation was unsatisfactory, but not because the description of phenomena using these concepts was deceiving, but rather precisely because these descriptions did accurately describe the phenomena and hence had to be taken at face value. Therefore, what was deceiving was the philosophical presupposition that implied that a choice had to be made: that a physical system as it exists independently of its phenomenal appearances had to *be* one or the other. For many physicists this wave-particle dualism may have indicated that wave and particle concepts must be abandoned, and new concepts must be developed for describing atomic phenomena. But Bohr saw that since in each single case theoretical predictions using the classical concepts did unambiguously describe the phenomena, there was absolutely nothing inconsistent in using wave and particle concepts in describing *phenomena*. Thus the "paradox" of wave–particle dualism was the result of the philosophical presupposition that the reason these classical pictures do so describe the phenomena is because the spatio-temporal parameters in terms of which these pictures are defined also refer to properties possessed by systems apart from any interaction in which they are observed. The retention of the classical concepts in quantum theory implies that this theory is inconsistent with any framework which presupposes that the use of terms to describe the phenomenal appearances of objects requires that these objects possess properties corresponding to these terms even when they are not in an observational interaction. If the classical framework makes such a presupposition, then so much the worse for that framework. But if the mathematical formalism provides a consistent and empirically confirmed representation, then, Bohr reasoned, that theory can be retained by reinterpreting its use of descriptive concepts in a new framework. This was Bohr's second great insight; it is the touchstone of all of his arguments in complementarity.

CHAPTER FOUR

THE BIRTH OF COMPLEMENTARITY

"The very nature of the quantum theory thus forces us to regard the space–time co-ordination and the claim of causality, the union of which characterizes the classical theories, as *complementary* but exclusive features of the description, symbolizing the idealization of observation and definition respectively."[1] In this sentence, containing implicitly many of the characteristics of Bohr's original formulation of complementarity, the word "complementary" appears for the first time in Bohr's published work. The statement appears in the published version of the talk given at Como. Bohr was invited there to the congress commemorating the death of Alessandro Volta "to give an account of the present state of the quantum theory in order to open a general discussion on this subject".[2] He used the occasion to present what he believed would be the final innovation completing the revolution which had begun with Planck's introduction of the quantum in 1900. Bohr hoped that his paper would resolve in a single stroke all the apparent inconsistencies ensuing from the attempt to work the quantum postulate into the classical framework. But he refrained from adopting a revolutionary tone, trying instead to "harmonize the apparently conflicting views taken by scientists".[3] Perhaps this is why the paper failed to impress its original audience. Nevertheless, in fact it extends Bohr's analysis of the philosophical foundations of natural science to new depths.

In spite of his conservative tone, it would surely be a mistake to assume that Bohr did not recognize the revolutionary character of his suggestions. Repeatedly he returned to the theme that once the quantum revolution had begun—once it was regarded as unavoidable that in order to account for an ever increasing range of phenomena, physics must accept the quantum postulate—there was no question of returning to the old order. The issue, then, at least for Bohr was how and in what form a new framework would reestablish harmony in the unsettled and divided state of physics. Hence Bohr's first presentation of complementarity looks both ways. Looking at the recent past, he saw complementarity as ending a period of turbulence during which physicists were reduced to working within the context of an inconsistent patchwork system, well characterized by Einstein's complaint to his friend Max Born that, "one ought really to be ashamed of the successes [of quantum theory], as they are obtained with the help of the Jesuitic rule: One hand must not know what the other hand does".[4] At the same time, looking to the future, Bohr hoped, with all of the might of his life's devotion to natural philosophy, that now physics had established the general outline of a new framework as consistent and inclusive as the classical framework had once been held to be.

This chapter considers the basic argument of complementarity *as Bohr originally presented it*; thus I shall concentrate on his presentation in the Como paper, which Bohr always considered to be his more or less fundamental statement. In the chapters following this one we shall deal with how his original view was refined, though never seriously altered, and extended to fields other than physics. But before discussing the explicit themes dealt with in the Como talk, it will aid our understanding to consider the genesis of the paper.

1. The Como Paper

By April of 1927 the differences with Heisenberg had been smoothed over, and Bohr began work, using Oskar Klein as sounding-board, on what was to be the Como paper. From early drafts we know that Bohr originally intended to write a brief letter to the Editor of *Nature* (he had contributed many such letters) commenting on an exchange of letters between Norman Campbell and Pascual Jordan, concerning Campbell's suggestion that time could be treated as a statistical variable. As it turned out, the letter to *Nature* was never completed, but it did serve as the seed from which the Como lecture grew.[5]

In 1921 Campbell had suggested that though quantum paradoxes were generated by the formal character of different "pictures" used to describe atomic structure, both "pictures" could refer to the *same* reality if the concept of "time" were modified. Though he found it "interesting", Bohr had countered Campbell's suggestion in a letter published in *Nature* on March 24, 1921. Campbell's idea that the "time of an individual event" is no more meaningful than the "temperature of an individual molecule", was unsupportable for a number of physical reasons; however, the course of the debate which ensued provided Jordan (who had worked with Heisenberg and Born to formulate matrix mechanics) with the opportunity to mention Heisenberg's uncertainty relations and thereby set the stage for discussing the use of space–time pictures in describing atomic phenomena. Jordan's letter appeared in *Nature* on May 28, 1927; so we may surmise that Bohr began writing in early June.

Klein recalls that "Bohr had been talking the whole winter about something which had, in some way, to do with complementarity. I don't think it was clear to him and certainly not to anyone else."[6] Although Bohr's ideas surfaced in the heated debates with Heisenberg, they were still too unformed to put into writing. Klein describes the first exasperating attempts as follows:

> The worst I had was in the summer of '27 working on complementarity. We were out in the country then, and I came every day to him, and he began to dictate. The next day that which was dictated was not used and we began over. ... I think he was really overworked, and of course the whole year before had been very strenuous. And then, in the spring, after Heisenberg's paper he had got his main ideas about complementarity; and then he tried to write them perhaps a little too early. Partly he was very tired and partly the ideas were not at the stage to be written, so I think there were several reasons why he did not succeed there.[7]

Thus in spite of a full summer's work, little progress was made. Klein notes that

> ... in the comparatively early spring—April or May—the main content was already there, but it was still in such a form that it was hard to connect it. He [Bohr] worked the whole summer ... during the summer nothing came of it, although an enormous number of pages were written. ... Only at the last moment, he was going to give a lecture at Como ... as he had promised. Then he and everybody around him were very upset.[8]

Nevertheless we can see in a letter to Heisenberg that by late August Bohr seemed relatively optimistic that he had formed his ideas into a cogent paper:

> In the course of the summer I have advanced a bit on the general physical problems and I think that I have succeeded in clarifying my thoughts to some degree. Since I have been a little tired I have not as yet managed to finish any article, but I hope in my lecture in Como, with which I am just now occupied, that I can give a fairly clear state-

ment of my views on the foundation of quantum theory. Of course in this lies essentially the same thing that we have talked about so much, but I find constant new instruction with stressing how even the aspects of unobservability, or better, undefinability, give the possibility for combining the demands of the individual single processes with the superposition principle. Also the close connection between space and time in relativity theory becomes very instructive from this viewpoint.[9]

Two points are worth noting in this letter. First, although the Como paper did not put the theme of continuity versus discontinuity at center stage of Bohr's arguments, the reference to "combining the demands of the individual single processes with the superposition principle" indicates that it was reflection on the alternative descriptions of particle and wave pictures which focused his attention on the question of definability and observability, so central to complementarity. Second, even at this time the connection between complementarity and the arguments concerning space and time in relativity theory was already present in Bohr's mind long *before* the struggle to win over Einstein had begun.

In spite of his optimism in this letter to Heisenberg, Bohr was definitely still ill at ease about how his new viewpoint would seem from the "outside", so to speak, for from the "inside" it appeared all so obvious. Thus five days later he wrote Darwin, who was also working on "the foundations of quantum theory":

Indeed I am dreadfully ashamed not yet to have finished any paper about the general views on the quantum theory about which we talked so often. I am afraid that I have been thinking so much over it, that sometimes *I am afraid it is all too trivial.* However, I feel that it has been growing steadily in my mind and I hope in Como to be able to give a reasonably clear account of my views.[10]

In fact, the lecture finally delivered in Como on September 16 was by no means "reasonably clear". It apparently had minimal effect on the audience and was "so short that nobody could understand it really".[11] One month later Bohr delivered the same address at the Solvay Conference in Brussels, again drawing essentially a negative response, this time including the criticism of Einstein. Consequently Bohr was compelled to spend the following winter, partially assisted by Pauli, in successive rewritings of the original talk. The final published copy of the paper, "The Quantum Postulate and the Recent Development of Atomic Theory", appeared in one version in the Congress Proceedings and in a substantially revised version in *Nature* in

1928. The latter version is also the centerpiece of the collection of Bohr's essays that was published in 1931 in German and in 1934 in English under the title, *Atomic Theory and the Description of Nature*. However, even with almost one full year's work on this paper, innumerable rewrites, two public deliveries, and three different sources of publication, the essay contains many obscurities and never makes very clear what this new "viewpoint" is supposed to be.

2. The Argument of Complementarity

Undoubtedly the nature of Bohr's task was unfamiliar to most physicists. Bohr himself was somewhat more used to dealing with issues on a purely conceptual level, as opposed to the formal mathematical approach more typical of the younger men. However, the paper suffers from divided attention; on one hand it surveys the series of developments and problems that characterize "the quantum revolution". But on the other hand it tries to lay the foundations for the framework of complementarity. It is not unlikely that its unenthusiastic reception was at least partially due to the fact that Bohr's audience was all too familiar with the former objective but all too unfamiliar with the latter.

As the title is meant to suggest, the Como paper purports to present an analysis of the consequences of accepting the quantum postulate. We have seen that the need for a new framework which would harmonize this postulate with the classical mode of description had haunted Bohr for nearly fifteen years, since he had first formulated the postulate as the basis for his atomic model of 1913. Now he intended to show how the repercussions of this assumption provide the primary grounds for his generalization of the classical framework.

Bohr presents virtually his entire argument in the six paragraphs which comprise the first section of the paper. Although it involves some repetition of quotations, Bohr's argument is so compressed that it is best presented paragraph by paragraph. This section of the published paper appears substantially the same in the earliest surviving typescript of Bohr's preliminary drafts of the paper.[12]

Bohr begins by calling attention to the "peculiar" difficulties which result from the quantum postulate:

> The quantum theory is characterized by the acknowledgment of a fundamental limitation in the classical physical ideas when applied to atomic phenomena. The situation thus created is of a peculiar nature, since our interpretation of the experimental material rests essentially on the classical concepts. Notwithstanding the difficulties which, hence, are involved in the formulation of the quantum theory, it seems, as we shall see, that its essence may be expressed in the so-called quantum postulate, which attributes to any atomic process an essential discontinuity, or rather individuality, completely foreign to the classical theories and symbolized by Planck's quantum of action.[13]

In this opening statement several important ideas already noted in the previous chapter appear. First Bohr is clearly looking for "a fundamental limitation" which will restrict the application of "the classical physical ideas when applied to atomic phenomena". As we have seen, once he recognized that wave–particle dualism was inescapable, he concentrated not on overthrowing these paradoxical representations but rather on removing the paradox by limiting their use. Second we should note that Bohr claims that these classical concepts must be retained because without them it would be impossible to provide what he calls an "interpretation of the experimental material". In other words, the classical concepts are necessary to describe the physical interactions which the theory treats as observations of atomic systems. Finally, we see the now familiar claim that the "difficulties" in formulating a quantum theoretical description of atomic phenomena are due to the fact that the "essence" of the theory, the quantum postulate, "attributes to any atomic processes an essential discontinuity, or rather individuality, completely foreign to the classical theories ...". Bohr's reference to this discontinuity as an "individuality" is his way of expressing the fact that because it has been discovered that it is necessary to describe interactions involving atomic systems with a theoretical formalism which represents them as taking place discontinuously, within that formalism it is impossible to define separately the classical state of each of the systems which interact. A state for the interacting whole system as an "individual" can be defined, but not for its separate components. Thus acceptance of the quantum postulate implies the impossibility of describing the observations in a way which subdivides the whole phenomenon of interaction into a process taking place between systems for which classical mechanical states can be precisely defined throughout the interaction.

In the next paragraph Bohr turns to consider the consequences of this "individuality" on the use of the classical concepts in the description of such a discontinuous interaction:

> This postulate implies a renunciation as regards the causal space time co-ordination of atomic processes. Indeed, our usual description of physical phenomena is based on the idea that the phenomena may be observed without disturbing them appreciably. This appears, for example, clearly in the theory of relativity, which has been so fruitful for the elucidation of classical theories. As emphasized by Einstein, every observation or measurement ultimately rests on the coincidence of two independent events at the same space–time point. Just these coincidences will not be affected by any differences which the space–time co-ordination of different observers otherwise may exhibit. Now the quantum postulate implies that any observation of atomic phenomena will involve an interaction with the agency of observation not to be neglected. Accordingly, an independent reality in the ordinary physical sense can neither be ascribed to the phenomena nor to the agencies of observation.[14]

Bohr's use of "phenomena" here is clearly intended to refer to the object of observation as distinguished in the description of the interaction from the "agencies of observation" which are described as the physical systems with which the "observed phenomena" interact in the observation. This use of "phenomena" differs from Bohr's later talk of the *whole* interaction as the "phenomenon" to be described by the theory.[15] The classical means of describing physical processes through "causal space time co-ordination" must be "renounced" because it presupposed that the observation necessary to determine "the coincidence of two independent events at the same space–time point" did not "disturb" the observed "phenomena" appreciably. However, the quantum postulate expresses the empirical discovery that this assumption is not true in describing observations of atomic phenomena.

In this early presentation of complementarity Bohr is still liable to the confusion of speaking of observation as "disturbing" the observed object. As we have noted in the previous chapter, the fact that an observation requires a physical interaction between the observed object and the observing system implies that observation may alter the state of an observed system, *i.e.*, that it "disturbs" it. This is already the case in classical mechanics. However, in that context, the presupposition that the observed and observing systems change state continuously in an interaction, allows one to determine theoretically the nature and extent of the disturbance; in Bohr's words, the interaction is "controllable". Thus classically it was possible on the basis of an observation to define the state of the closed system isolated from interaction by correcting for the disturbance produced by observing it. In the classical framework, although determining the state of an isolated system directly by observation is represented as physically impossible, the classical mechanical state could be *defined* on the basis of observations because the disturbance produced by observation either could be considered negligible or was controllable.

Unfortunately, speaking this way tended to confuse Bohr's audience into thinking that he was claiming that in fact the systems described by quantum theory do exist in classical mechanical states while in an observational interaction, but because the observation "disturbs" those states it is impossible to determine empirically precisely what they are. On *this* interpretation, Bohr's talk of a "limitation" on the classical descriptive concepts would appear to refer to a limitation on *knowledge* of the classical mechanical states. That such a reading of Bohr's words is *not* what he intended is clearly seen by the fact that it is inconsistent with Bohr's primary conclusion, expressed in the last sentence. There he concludes that we cannot describe an observation in a way which treats the observed object and the observing instrument as having "an independent reality in the ordinary physical sense", *i.e.*, in the sense of the "waves" and "particles" of classical physics, because if the objects described by quantum mechanics as such waves and particles did in fact have "independent reality in the ordinary physical sense", it would be possible to define classical mechanical states for them. Since by the quantum theory this is not possible, if the theory provides a complete description, these objects cannot be regarded as existing in classical mechanical states whether or not it would be possible for us to determine these states. If Bohr had believed that the observed object does exist in a well-defined classical mechanical state, though we cannot determine it because of the alleged "disturbance", there would be no reason why we could not continue to regard the observed object as having "independent reality in the ordinary physical sense". But as we shall see in the following chapters, with respect to the nature of physical reality, the fact that we can no longer so regard the observed object presents the most dramatic break with the classical framework that is precipitated by accepting the quantum postulate.

The disturbance language which Bohr uses at the beginning of this paragraph tends to lead the reader away from this important conclusion. However, a second line of argument, derived from the "individuality" of the observational interaction, does not easily lend itself to this misinterpretation. This argument which is more prominent in later essays is already evident in the following sentences which were appended to the above paragraph only in later versions of the manuscript:

> After all, the concept of observation is in so far arbitrary as it depends upon which objects are included in the system to be observed. Ultimately, every observation can, of course, be reduced to our sense perceptions. The circumstance, however, that in interpreting observations use has always to be made of theoretical notions entails that for every particular case it is a question of convenience at which point the concept of observation involving the quantum postulate with its inherent "irrationality" is brought in.[16]

This passage presents a distinct way of regarding the observational interaction not liable to the misunderstanding involved in the "disturbance" interpretation. After Bohr explicitly repudiated the "disturbance" way of speaking, he emphasized ever more the fact that the process of interaction could not be non-arbitrarily subdivided.[17] Because of the individuality of the interaction described as an observation, any distinction between observed object and "agency of observation" is "arbitrary". Since it is arbitrary in any interaction what we wish to consider the object and what the observing system, it is "a question of convenience" at what point we wish to make the discontinuity of change of state required by the quantum postulate enter the description of the interaction.

Bohr's talk of "interpreting observations" refers to the need to describe the observation as an interaction between physical systems. Any such description must be made within the context of a conceptual framework which represents such interactions as, say, a collision between particles or the interference effect produced by a superposition of different waves in the field, etc. Hence such an "interpretation" must make use of "theoretical notions". Since the discontinuity in change of state occurs in the description of the interaction between observed object and observing system, wherever we decide to make that distinction in the whole interaction is the point at which the discontinuity will enter the description. And that decision is an "arbitrary" matter of convenience. Bohr intended the reference to this quantum discontinuity as an "irrationality" to mean that the existence of such a discontinuity in nature is inconsistent with the classical framework. In no case did he mean that "reason" was to be abandoned in describing quantum processes, as in some cases this unfortunate choice of terms was wildly misinterpreted as intending.

In the next paragraph Bohr turns to consider the consequence of this individuality of quantum processes on the classical ideal of description:

> This situation has far-reaching consequences. On the one hand, the definition of the state of a physical system, as ordinarily understood,

> claims the elimination of all external disturbances. But in that case, according to the quantum postulate, any observation will be impossible, and above all, the concepts of space and time will lose their immediate sense. On the other hand, if in order to make an observation possible we permit certain interactions with suitable agencies of measurement, not belonging to the system, an unambiguous definition of the state of the system is naturally no longer possible, and there can be no question of causality in the ordinary sense of the word.[18]

Here we see that Bohr uses "definition" to refer to the classical goal of defining the state of the isolated system, from which all "external disturbances" have been "eliminated". He points out that this goal, necessary for applying the conservation principles to determine the causal development of the state of the system is impossible to achieve simultaneously with giving the space and time concepts empirical reference ("immediate sense") in an observation. If an observation is made, the observed system is in an interaction, and in the quantum representation the "individuality" of this interaction means that any division between observing system and observed object is an arbitrary one made within the *description* of that interaction. Thus it is impossible to formulate an "unambiguous" definition of the system's classical mechanical state. Without such a definition of the closed system's state, it is impossible to apply the conservation principles (the "claim of causality") to the state of the isolated system to provide a *causal* description of the temporal development of that state. The mode of space–time coordination is necessary to describe what is determined by observation, and it has also been empirically determined that the conservation principles do apply to each atomic process. Hence although the two classical modes of description cannot be applied at the same time to the same physical system, neither can they be abandoned. Thus in the next sentence, already quoted at the beginning of this chapter, Bohr concludes that space–time coordination and causal description are "complementary":

> The very nature of quantum theory thus forces us to regard the space–time co-ordination and the claim of causality, the union of which characterizes the classical theories, as complementary but exclusive features of the description, symbolizing the idealization of observation and definition respectively. Just as the relativity theory has taught us that the convenience of distinguishing sharply between space and time rests solely on the smallness of the velocities ordinarily met with compared to the velocity of light, we learn from the quantum theory that the appropriateness of our usual space–time description depends entirely on the small value of the quantum of action compared to the actions in-

> volved in ordinary sense perceptions. Indeed, in the description of atomic phenomena, the quantum postulate presents us with the task of developing a "complementarity" theory the consistency of which can be judged only by weighing the possibilities of definition and observation.[19]

Here it is clear that in this original statement of Bohr's new viewpoint, the complementary relationship holds between space–time co-ordination and the claim of causality, both of which were combined in the classical framework. Only later does Bohr speak of complementarity between wave and particle "pictures". Bohr's initial concern is to explain why these two modes of description could have been united classically and why they no longer can be so combined in describing atomic phenomena. The reason he gives is because processes in the classical framework normally could be described on the basis of observations involving "ordinary sense perceptions" which require interactions so huge relative to that measured by Planck's quantum, that for all practical purposes the discontinuity in the interaction could be ignored. This reasoning is an instance of Bohr's correspondence principle at work. Furthermore, it provides his warrant for regarding the classical framework as a special case of which complementarity is a generalization. To underscore his point, Bohr analogizes the lesson of quantum theory to that of relativity theory, a favorite way of making his argument. Had our everyday world involved large velocities relative to the speed of light, the classical Newtonian distinction between space and time could not have been made. So, similarly, had our everyday experience involved interactions in which the quantum of action loomed large, the classical union of causal and space–time modes of description would not have been possible.

Although the two modes of description are not able to be applied simultaneously to the same object, a consistent use of the two in a complementary fashion is possible because those situations which allow the application of the conservation principles to define the state of a system isolated from interaction are mutually exclusive with those situations which allow the application of the space and time concepts to describe the system as observed. By "weighing the possibilities of definition and observation" Bohr and Heisenberg had not been able to discover any physical interaction which permitted *both* forming a space–time picture of the system *and* applying the conservation principles to determine its causal change through time. Consequently Bohr believed that no inconsistency could result from the complementary use of both modes of description. However, he leaves open an avenue for possible refutation of his complementarity viewpoint: if an experimental interaction with a quantum mechanical system can be found which permits both a space–time coordination of the results of observation and the use of the conservation principles to define the system's state isolated from the interaction, then Bohr's claim that these two modes of description are mutually exclusive but complementary would be seen to be incorrect. As we shall see in the next chapter, this is an approach which Bohr's critics did in fact pursue.

The next paragraph begins with a transition from the relation between these modes of description in classical and complementarity frameworks to the consequences of this change of frameworks for "the much discussed question of the nature of light and the ultimate constituents of matter".[20] The remainder of the paragraph focuses on the nature of light, here used as an example of all radiative phenomena. Bohr calls to mind both the phenomena which require wave descriptions and those which require particle descriptions. Alluding to his own revolutionary efforts, he reminds the reader that this "apparent contradiction" cannot be overcome by attempting to discard the strict applicability of the conservation principles, for this attempt has "been definitely disproved through direct experiments".[21] He then turns to consider the implications of his viewpoint for particle and wave descriptions:

> This situation would seem clearly to indicate the impossibility of a causal space–time description of the light phenomena. On the one hand, in attempting to trace the laws of the time–spatial propagation of light according to the quantum postulate, we are confined to statistical considerations. On the other hand, the fulfillment of the claim of causality for the individual light processes, characterized by the quantum of action, entails a renunciation as regards space–time description. Of course, there can be no question of a quite independent application of the ideas of space and time and of causality. The two views of the nature of light are rather to be considered as different attempts at an interpretation of the experimental evidence in which the limitation of the classical concepts is expressed in complementary ways.[22]

Here Bohr argues that "we are confined to statistical considerations" in attempting a space–time description of light propagation because the quantum formalism does not permit determining the state of the observed radiation in an observational interaction with any greater precision than that expressed by Heisenberg's uncertainty relations. Thus if we have an observation of a phenomenon which is interpreted by representing radiation as a photon at a precise point in space and time, the energy of that photon cannot be defined, for it requires determining the frequency of the radiation, and that in turn requires a wave description which represents the radiation as spread out through a region of space. An energy or momentum observation will introduce a parallel indeterminacy into the representation of space–time position. In either case the lack of determination in defining the state of the observed object implies that the use of the conservation principles to determine its future state, will yield only statistical predictions. The "two views of the nature of light", wave and particle representations, permit "different attempts at interpretation of the experimental evidence" in the sense that they permit describing the interaction in different ways so as to allow either a space–time co-ordination or a momentum–energy determination, but their application is limited "in complementary ways" as expressed by the uncertainty relations.

Having considered the dualism of wave and particle descriptions in the case of electromagnetic radiation, in the second to last paragraph Bohr turns to the "analogous situation" with respect to the "nature of the constituents of matter". As before, he points to the experimental evidence concerning various phenomena which can be described only by representing the electron as a particle in some situations and as a wave in others. He continues by making the same point as in the case of light:

> Just as in the case of light, we have consequently in the question of the nature of matter, as far as we adhere to classical concepts, to face an inevitable dilemma which has to be regarded as the very expression of experimental evidence. In fact, here again we are not dealing with contradictory but complementary pictures of phenomena, which only together offer a natural generalization of the classical mode of description.[23]

The use of the two classical pictures of waves and particles to represent both the constituents of the atomic system and light is the logical consequence of the complementary application of the two classical modes of description. Since these pictures appear only in complementary descriptions, Bohr speaks of them as "complementary pictures of the phenomena". Since the classical modes can no longer be simultaneously applied to describe an observed object, there is no possibility of representing the observed object as *simultaneously* both a particle and a wave.

However, in "interpreting" an observation, we must use "theoretical notions" to describe the observed object both before and after the actual observational interaction. Thus we must use these same particle and wave concepts to represent theoretically the object before or after the interaction as though they referred to an independently real physical object isolated from any observational interaction. As we have already seen, such pictures represent objects as classical systems existing in well defined states, but the theoretical formalism prohibits defining the observed system as existing in such a state. This fact implies that these pictures cannot be used to ascribe "an independent reality in the ordinary physical sense" to the objects they are used to represent. Thus Bohr concludes the above paragraph with the important claim that while the use of wave and particle pictures to represent the quantum mechanical object apart from the observational interactions in which we observe this object is necessary for a theoretical representation of an observational interaction, these "pictures" refer to "abstractions":

> In the discussion of these questions, it must be kept in mind that, according to the view taken above, radiation in free space as well as isolated material particles are abstractions, their properties on the quantum theory being definable and observable only through their interactions with other systems. Nevertheless, these abstractions are, as we shall see, indispensable for a description of experience in connection with our ordinary space–time view.[24]

Using wave and particle pictures, we can, of course, formulate theoretical representations of the objects of quantum mechanical descriptions as independent from an observational interaction, *i.e.*, as "radiation in free space" or as "isolated material particles". Indeed, Bohr makes it clear that "a description of experience", (*i.e.*, "interpreting" an observational interaction) makes such a representation "indispensable". But the apparently paradoxical nature of the use of particle representations in some situations and wave representations in others is dispelled once it is understood that such descriptions are "abstractions" (elsewhere Bohr also refers to these pictures as "idealizations") in the sense of purely theoretical representations, not descriptions of the properties of independently real entities existing apart from observation. This conclusion is the very core of the difference between how the description of nature is understood within the classical framework and how it is understood within complementarity.

The last paragraph of the first section of the Como paper marks the transition to the remainder of the paper in which Bohr turns his attention to showing how the uncertainty relations express the complementary aspects of the description. We shall return to consider this matter in the last section of this chapter, but before doing so, it will be helpful to discuss at greater length some of the points Bohr makes in the argument of this section.

3. Comments on Bohr's Original Argument

The argument presented in the Como paper always remained Bohr's basic approach to complementarity. He expanded on this formulation in two subsequent essays both published in 1929, but neither include any specifically new points bearing on his argument or any substantive revision in his terminology. These papers with the Como paper and the 1925 paper on "Atomic Theory and Mechanics" were published with an "Introductory Survey" in *Atomic Theory and the Description of Nature.* This slim volume of only 119 pages, written without any overall structure provides Bohr's fundamental statement of the viewpoint of complementarity. However, in the more than twenty-five other articles bearing on complementarity that Bohr wrote between 1931 and his death in 1962 there are included substantial changes in both mode of expression and emphasis, as well as considerable further comment on the application of complementarity to fields beyond atomic physics. We will consider the most important changes in expression and emphasis in the following chapter and then turn to the extension of complementarity in Chapter Six. However, in the present section, it will help our understanding of complementarity to consider some additional points inherent in his original formulation of the argument.

Throughout the previous chapter we noted how strongly Bohr stresses that the full range of observed phenomena requires that both matter and radiation be described through the seemingly contradictory pictures of waves and particles. The paradoxical nature of this dualism is due to the classical tendency to regard these pictures as representations of objects having an "independent reality", the properties of which correspond to the properties of the "visualizable pictures". The justification for this position lies in the classical description of interaction as taking place continuously, which makes it possible to define the state of a system at any point in its transition from one state to another. The wave and particle pictures derived from the classical definition of the state of the system thus enabled one to "visualize" that system throughout the course of its interaction with observing instruments.

Once the change of state in an interaction is represented as taking place discontinuously, it is not possible to provide a spatio-temporal picture of the system changing from one state to another. Thus in the interaction the state of the observed system cannot be defined separately from that of the observing system. This gives the interaction the feature of indivisibility which Bohr calls "individuality". Consequently it is impossible to visualize the interaction as a "transition process":

> Indeed only by a conscious resignation of our usual [*i.e.*, classical] demands for visualization and causality was it possible to make Planck's discovery fruitful in explaining the properties of the elements on the basis of our knowledge of the building stones of atoms. Taking the indivisibility of the quantum of action as a starting-point, the author suggested that every change in the state of an atom should be regarded as an individual process, incapable of more detailed description, by which the atom goes over from one so-called stationary state into another. ... On the whole, this point of view offers a consistent way of ordering the experimental data, but the consistency is admittedly only achieved by the renunciation of all attempts to obtain a detailed description of the individual transition process.[25]

This "individuality" of the interaction, makes it impossible to define a classical state for the atomic system as it is observed in an interaction and thus to uphold the classical justification for according an independent reality to the objects visualized through the wave and particle pictures of the interaction. But the tendency to accord an independent reality to observed objects even in the quantum theoretical representation persists as long as we see the description of nature from the classical viewpoint.

In the light of Bohr's concern with this paradoxical dualism, it is perhaps remarkable that the relation between wave and particle descriptions is *not* the basis for Bohr's introduction of the word "complementarity". It is clear that Bohr uses "complementarity" in the Como paper to refer to the relationship between space–time description and causal description. As we shall see in the next chapter, this way of speaking was replaced with his favorite reference to the "complementarity" existing between "phenomena". But, this talk of "complementary phenomena" is perfectly consistent with holding that the basic complementary relationship holds between space–time coordination and applying the claim of causality. Even in his latest essays this is the way in which Bohr thought of his viewpoint.[26]

Nevertheless, as early as the Como paper, and throughout his other essays, Bohr also refers to complementarity between wave and particle "pictures". Indeed, complementarity was designed to remove the paradoxical quality of this dualism by providing an alternative to the classical framework. Thus he holds that "the well known dilemma between the corpuscular and undulatory character of light and matter [is] avoidable only by means of the viewpoint of complementarity".[27] In the Como paper, referring to the relation between "corpuscles" and de Broglie's wave representation of "the constituents of matter", Bohr remarks that, "... we are not dealing with contradictory but with complementary pictures of the phenomena, which only together offer a natural generalization of the classical mode of description".[28] Referring to the same dualism with respect to the nature of "light phenomena", he provides a more enlightening remark: "The two views of the nature of light are rather to be considered as different attempts at *an interpretation of experimental evidence* in which the limitation of the classical concepts is expressed in complementary ways."[29]

Unfortunately, beyond these brief comments, Bohr never directly explains the relationship between the complementarity of space–time and causal modes of description and the complementarity between wave and particle pictures. Nevertheless, it is possible to account for this relation. Bohr cannot mean to associate space–time description with "particle pictures" and causal description with "wave pictures". The unambiguous use of the term "particle" refers to that which exists in a classical mechanical state, and this concept of the classical state requires determining *both* the position and the momentum of the particle. Thus any well-defined use of "particle" cannot be the result of applying only the mode of space–time description. The same fact holds true for the use of the concept of a wave in a field.

However, empirical science rests on observations, so an adequate theory must provide an account of observation. In physical science this is done by treating an observation as an interaction between physical systems. This is particularly obvious in atomic physics where the observing system is never a simple human sense organ, as it can be in classical physics, but is always a piece of laboratory equipment designed specifically for the purpose of exhibiting the sorts of phenomena which are described as revealing the behavior of atomic systems. These interactions *are* the observations which allow the complementary modes of description to determine the spatio-temporal locus *or* the energy and momentum of the objects with which the measuring instruments interact.

As we saw in the previous chapter, the problem of describing interactions between radiation and material particles became the major problem for quantum theory in Bohr's mind. The empirical discovery of wave–particle dualism convinced him that wave and particle concepts cannot be regarded as picturing an independent physical reality, but these pictures are indispensable, because they provide the only way for interpreting the observational interactions by unambiguously distinguishing the measuring system from the atomic systems. The very concept of observation implies that a distinction between observing agency and observed object must be made. So in order to describe the interaction as an "observation", there must be concepts for distinguishing between the interacting systems. Each interaction must be "interpreted", in Bohr's words, in order to describe how the measuring instruments determine the property of the observed object. In such a theoretical interpretation of the observation, each system must be unambiguously represented so that the interaction can be described in a way which characterizes the observed object in terms of *either* spatio-temporal concepts *or* the "dynamical" concepts of momentum or energy. At this point the classical "wave" and "particle" pictures have their necessary use, for through them the interaction can be described as one which determines the phenomenal properties to which the classical mechanical terms refer. However, in order to apply the space–time mode of description, an interaction with the system which is to be described must occur which physically precludes the necessary interaction to apply the conservation principles in the causal mode of description to determine the "dynamical behavior" of the system:

> ... since the discovery of the quantum of action, we know that the classical ideal cannot be attained in the description of atomic phenomena. In particular, any attempt at an ordering in space–time leads to a break in the causal chain, since such an attempt is bound up with an essential exchange of momentum and energy between the individuals and the measuring rods and clocks used for observation; and just this exchange cannot be taken into account if the measuring instruments are to fulfill their purpose. Conversely, any conclusion, based in an unambiguous manner upon the strict conservation of energy and momentum, with regard to the dynamical behavior of the individual units obviously necessitates a complete renunciation of following their course in space and time.[30]

Thus making a space–time measurement requires describing a different interaction from that required to determine energy or momentum. Since those interactions which can be described as spatio-temporal observations will differ from those interactions which are described as momentum–energy observations, *different pictures* may be used to describe the observed object in these different observations without inconsistency or contradiction. Since the interactions allow the application of the complementary modes of description, by association we may refer to the pictures in terms of which these interactions are described as "complementary pictures". However, in doing so, we must be careful not to attempt any one-to-one correlation of one type of picture with one or the other descriptive mode.

In the space–time mode of description, particle and wave pictures permit an account of the observational interaction which allows the object to be pictured as physically distinct from the observing instruments, but in order to determine the causal effect of this interaction on the measuring system, it is necessary to employ the causal mode of description using the conservation principles. When we do this, the individuality of the interaction makes it impossible to define simultaneously the spatio-temporal picture of the interacting systems. Consequently we cannot form an exact space–time picture of the interacting systems *and* at the same time describe the causal effect one system has on the other. Thus in order to "interpret" an observation, the interacting systems must be clearly distinguished through using the space–time mode of description. But to describe the effect one system has on the other, the causal mode is necessary. This conclusion justifies regarding the two modes of description as "complementary" rather that merely alternative ways of describing nature. Furthermore, it also implies that neither mode of description can be abandoned, but neither can they be used in conjunction, as in the classical framework.

Since the classical definition of a mechanical state requires determining the simultaneous position and momentum or energy and time parameters characterizing the observed object, the ideal of defining the *classical* state of an isolated system now appears to be unattainable. However, in order to use the conservation principles to interpret the causal effect of the interaction, the state of the object before and after the interaction must be capable of being defined. Thus, the quantum mechanical formalism must provide an unambiguous means to represent theoretically the state of the systems when they are *not* observed, *i.e.*, as "isolated" physical systems. This fact makes the "abstractions" of "radiation in free space" and "isolated material particles", in Bohr's words, "indispensable" for the interpretation of an observation. But since the theoretical definition of the state of an isolated system *in the quantum formalism* cannot represent the system in terms of simultaneous position and momentum or energy and time parameters, it is impossible to derive any well-defined wave or particle "picture" of the isolated system from such a formal "abstraction". Though the quantum postulate implies that we cannot accord an independent reality to wave and particle pictures of an isolated system, the quantum theoretical representation of the system as isolated from interaction provided by the formalism provides a theoretical abstraction which is necessary to apply the conservation principles to account for the effect of an observational interaction on the observing system. As Bohr points out, although we must use wave and particle pictures to "interpret" observations, the classical expectation engendered by the use of these concepts that we can "visualize" the object of observation is no longer possible in quantum theoretical descriptions, for once the quantum postulate is accepted, the state of the system while observed cannot be defined separately from the instruments with which it is interacting and the state of the system as isolated is an abstraction. All we need from such theoretical representations of the isolated system is the "inner consistency" of a formalism which correctly predicts what will be observed. Therefore the classical expectation to visualize an independent reality is neither reasonable nor necessary.

The preceding discussion explains the relation of wave and particle pictures to complementary modes of description, but it raises the further question of why the classical concepts must be used at all. Indeed, it may seem obvious that the proper inference to draw from the fact that space–time and causal modes cannot be simultaneously applied is that these modes of description need to be replaced with some new alternative mode of description which would eliminate wave–particle dualism and restore determinism. However, Bohr clearly rejected any search for such an alternative:

> ... the view has been expressed from various sides that some future more radical departure in our mode of description from the concepts adapted to our daily experience would perhaps make it possible to preserve the ideal of causality also in the field of atomic physics. Such an opinion would, however, seem to be due to a misapprehension of the situation. For the requirement of communicability of the circumstances and results of experiments implies that we can speak of well-defined experiences only within the framework of ordinary concepts. In particular it should not be forgotten that the concept of causality underlies the very interpretation of each result of experiment, and that even in the coordination of experience, one can never, in the nature of things, have to do with well-defined breaks in the causal chain. The renunciation of the ideal of causality in atomic physics which has been forced on us is founded logically only on our not being any longer in a position to speak of the autonomous behavior of a physical object, due to the unavoidable interaction between the object and the measuring instruments which in principle cannot be taken into account, if these instruments according to their purpose shall allow the unambiguous use of the concepts necessary for the description of experience. In the last resort an artificial word like "complementarity" which does not belong to our daily concepts serves only briefly to remind us of the epistemological situation here encountered, which at least in physics is of an entirely novel character.[31]

Bohr's rejection of any search for a new set of descriptive concepts which would avoid wave–particle dualism and the limitations expressed by the uncertainty principle is a consequence of his conviction "that we can speak of well-defined experiences only within the framework of ordinary concepts". Elsewhere he makes the same point that the "basic concepts of mechanics" are "indispensable ... for the definition of fundamental properties of the agencies with which they [the atomic systems] react".[32] Thus to report an unambiguous description of the results of an experiment we must "interpret" what we observe as the results of an interaction between the measuring instruments and the atomic system which is "observed" in that interaction: "... our knowledge of the properties of the atomic particles ... is derived from measurements interpreted on classical mechanical lines."[33] It is clear that Bohr regards an "observation" of an atomic system as a *physical* process in which there is no need to appeal directly to the experiences of a conscious subject. However, a physical interaction serves as such an observation in atomic physics only if the observing instruments can be described unambiguously as they are "experienced" by the experimenter. The framework of classical physics has been developed for precisely such a task; in this respect, there is no lack of determination in the application of these concepts to describe the results of the observation which some putative alternative framework could provide.

In Bohr's later essays, the necessity for the classical concepts to communicate unambiguously the "circumstances and results of experiments" becomes the basis for his claim that complementarity provides an "objective description" of the phenomena with which atomic physics is concerned. But since Bohr's concern with objectivity is absent from his earlier essays, I shall postpone examining it until Chapter Seven. My present point is to examine Bohr's view that the necessity for the classical concepts lies in the very nature of an observation or "measurement" as providing the empirical basis for the theory:

> ... by an experiment we simply understand an event about which we are able in an unambiguous way to state the conditions necessary for the reproduction of the phenomena. In the account of these conditions, there can, therefore, be no question of departing from the Newtonian way of description and, in particular, it may be stressed that by the clocks which together with the scales are used to define the space–time frame, we simply understand some piece of machinery as regards the working of which classical mechanics can be entirely relied upon and where, consequently, all quantum effects have to be disregarded.[34]

The experiences on the basis of which the experimenter describes the observing instruments are experiences of everyday objects in space and time, thus space–time concepts refer directly to the experience of the observing instruments, enabling unambiguous description of the experiment. However, the use of the classical concept of causality is also essential, "for it must not be forgotten that the concept of causality underlies the very interpretation of each result of experiment, and that even in the coordination of experience one can never, in the nature of things, have to do with well-defined breaks in the causal chain".[35] Thus without the concept of causality it would be impossible to consider the state of the observing systems as an effect of its interaction with the observed object.

Bohr's view that the causal mode of description is necessary to interpret observation may contrast with the indeterminism of quantum theoretical predictions. The fact that the individuality of the interacting systems prevents defining the state of the system when it is observed implies that the causal description must renounce the classical ideal of deterministic predictions. In the traditional framework such predictions were possible only because the observed system could be described as in some determinate mechanical state throughout the observation. This can no longer be done, and we cannot use the abstraction of the quantum theoretical representation of the state of the isolated system to make deterministic predictions precisely because it is subject to the limitations of the uncertainty principle. The fact that a measurement must be described through the classical mechanical concepts conjoined with the fact that the formalism does not permit a well-defined classical state for the observed object, implies that if "causality" is identified with determinism, then "the invocation of the classical ideas, necessitated by the very nature of measurement, is, beforehand, tantamount to a renunciation of a strictly causal description".[36] However, the loss of determinism in quantum physics "does not imply that the laws of conservation of momentum and energy loose their validity, but only that their application stands in an exclusive, so-called 'complementary', relationship to the analysis of the motion of the particles".[37] Thus talk about the "renunciation" of causality in atomic physics should not mislead one into thinking that the classical account of causal reactions between systems is abandoned. Such an abandonment would preclude describing interactions as observations of the atomic systems with which the observing agencies interact. Bohr's point is that this means of describing interactions is *complementary* to giving a space–time description of such atomic systems. Indeed, although it is not a typical way to present complementarity, Bohr remarks that "space–time co-ordination and dynamical conservation laws may be considered as *two complementary aspects of ordinary causality* which in this field [quantum physics] exclude one another to a certain extent, although neither of them has lost its intrinsic validity".[38]

Bohr's recurrent claims that we must "renounce the customary demand for visualization" in describing atomic systems may seem inconsistent with his view that in complementarity observed phenomena must be described in terms of the ordinary space–time concepts of the classical framework. Bohr has no objection to the heuristic value of visualization. Nevertheless, from the very beginning he was somewhat upset over the tendency of many physicists to take his model too concretely, and occasionally he was himself drawn into taking it more literally than he had intended. If we cannot use the space–time concepts in their "usual" classical roles, and if we accept the essential correctness of the quantum mechanical formalism, as Bohr certainly did, then the classical ideal expressed in terms of "visualizing" atomic systems through wave and particle pictures must be rejected in favor of understanding the description of nature in a new way.

Bohr's choice to revise our understanding of the use of these descriptive concepts enables complementarity to hold that a theoretical representation of an isolated system is an abstraction from which one can make predictions expressed in terms of the spatio-temporal parameters which characterize the observed object *or* in terms of parameters defining its dynamical behavior, as one must in order to confirm the theory by observation. But at the same time, complementarity does not assume that these phenomenal properties which confirm the theoretical representation are the causal effects of corresponding properties possessed by an independent reality existing apart from the observational interactions necessary to give empirical reference to the spatio-temporal descriptive terms. Thus, for example, the reason the position of a particle or the frequency of a wave can be measured is not to be attributed to the presumed fact that "real particles" objectively trace paths through a real space–time, or that a real wave with a particular frequency is moving through free space that, in some way beyond empirical confirmation, corresponds to the space and time of observed phenomena. Such a correspondence between phenomena and an alleged independent reality must remain forever beyond the possibility of empirical investigation. To be sure, we can and must measure such empirical spatio-temporal observables in empirical science, but not because there are corresponding properties possessed by independently real entities which interact with measuring agencies to produce these observed phenomena. According to complementarity, the reason we *must* use such spatio-temporal concepts is because the way in which the theoretical representation is confirmed as adequate necessitates that at least some concepts find consistent and unambiguous empirical reference in the description of the physical systems which are treated as the observing apparatus. The classical viewpoint justified its application of spatio-temporal concepts to both the properties of observed phenomena and the properties of an independent physical reality only by assuming that the classical concept of the state of an isolated system is *not* simply an abstraction on which to base a theoretical prediction of the change of state caused by an observational interaction, but pictures the properties of an independent reality. However, once the quantum postulate is accepted, the quantum theoretical representation of an isolated system refers to an abstraction, and wave and particle pictures used to interpret an observation can no longer be regarded as describing an independent reality.

4. Complementarity and the Uncertainty Principle

The third section of the Como paper discusses how complementarity resolves the apparently paradoxical nature of the uncertainty relations. Since these relations express the crux of the non-classical aspect of quantum theory, understanding them through complementarity will advance our appreciation of Bohr's viewpoint.

Bohr's analysis was essentially coeval with Heisenberg's discovery, but he never intended complementarity as *merely* an explanation of the uncertainty principle. Einstein's criticisms forced him repeatedly into an analysis of experimental situations designed to overcome the limits of the principle; thus in the years following the Como paper, Bohr's original approach from an analysis of the application of the classical modes of description was largely lost from sight.

The belief that Bohr designed complementarity simply as his analysis of the uncertainty principle is historically unfounded, but it is a natural mistake, for he is anxious to show how complementarity explains this principle. For physicists raised in the framework of classical physics, the discovery of the uncertainty principle was a stunning surprise, thus it is hardly remarkable that in speaking to a group of physicists at Como Bohr was very eager to show how his new framework was able to interpret this principle. Thus, although his arguments do not make Heisenberg's discovery a *premise* of complementarity, it is not surprising that his analysis of it forms a significant portion of his early presentations of complementarity.

As Bohr understood the matter, both his new framework and Heisenberg's principle were the consequences of the quantum postulate; his was the consequence for the conceptual framework within which phenomena are described, while Heisenberg's discovery was its formal, mathematical consequence. In the mathematical formalism of classical mechanics, the two canonically conjugate parameters necessary to define the state of a system are independent of each other. However, in the quantum mechanical formalism these parameters are not independent but are linked reciprocally by the measure of discontinuity in change of state symbolized by Planck's quantum. Thus the uncertainty principle is the mathematical expression of the fact that in the quantum mechanical formalism, the classical ideal of a causal space–time mode of description made possible by defining the state of the isolated system is unattainable. This discovery then suggests either the need for accepting a new ideal for describing physical systems in the atomic domain and hence a new framework (Bohr's view), or the fact that the quantum mechanical description is "incomplete" (Einstein's position).

It follows mathematically from the uncertainty relations that the classical state parameters of the system can be defined only within the limits fixed by the element of discontinuity involved in the change of state of a quantum mechanical system. Therefore, since the initial conditions defining this state cannot be defined beyond this limit, theoretical predictions of the future state of the system are "indeterministic" or probabilistic in character. However, although the principle is logically entailed by the quantum mechanical formalism, the formalism itself cannot tell us whether this limitation is merely a limit on knowledge or reflects a deeper ontological "indeterminacy". In either case, accepting the prediction of a statistical probability as the maximum determination towards which the description of the causal development of the system can aspire has seemed repugnant to many physicists because classical mechanics allowed predictions which in principle could achieve any arbitrary degree of determinateness. Nevertheless, this classical strict determinism may be regarded as a special case of statistical determinism in which the statistical spread over the observable in question is negligible relative to the precision of the measurement. Consequently the classical ideal of strict determinism is possible in predictions involving interactions where the degree of discontinuity relative to the whole interaction can be ignored. Since these are in fact the sort of interactions which are adequately described classically, in line with the expectations engendered by Bohr's correspondence principle, the predictions of quantum theory will approach those of classical mechanics as the exchange of energy or momentum involved in an interaction increases in size relative to that expressed by Planck's constant. In other words, as we deal with larger and larger interactions, statistical determinism will converge into strict determinism.

The statistical character of predictions from the quantum mechanical formalism might be interpreted as implying that such predictions describe the behavior of large collections of atomic systems, each of which can be described by as yet undiscovered deterministic laws. Indeed, the statistical spread for the values of the classical parameters predicted by the formalism can be confirmed only by measurements on many atomic systems. However, Bohr categorically rejects this interpretation:

> One might perhaps believe that the properties of the elements do not inform us directly of the behavior of individual atoms but, rather, that we are always concerned only with statistical regularities holding for the average conditions of a large number of atoms. ... The elements have, however, other properties which permit of more direct conclusions being drawn with respect to the states of motion of the atomic constituents. Above all, we must assume that the quality of the light which the elements in certain circumstances emit and which is characteristic of each element is essentially determined by what occurs in a single atom.[39]

It will be recalled that Bohr proposed the quantum postulate to develop a mechanics that would represent the atomic system as mechanically stable. This property of stability must be attributed to each individual atomic system; thus the quantum postulate expresses a fact about the behavior of individual atoms. Since this postulate entails the statistical nature of the predictions of the formalism, the loss of strict determinism must be a consequence of the behavior of individual atoms. Thus Bohr declares as "vain" the ...

> ... repeatedly expressed hopes of avoiding the essentially statistical character of quantum mechanical description by the assumption of some causal mechanism underlying the atomic phenomena and hitherto inaccessible to observation. ... Above all such hopes would seem to rest upon an underestimate of the fundamental differences between the laws with which we are concerned in atomic physics and the everyday experiences which are comprehended so completely by the ideas of classical physics. ... the peculiar stability properties of atomic structures which are in obvious contrast with the properties of any mechanical model, but which are so intrinsically connected with the existence of the quantum of action, form the very condition for the existence of the objects and measuring instruments, with the behavior of which classical physics is concerned.[40]

The difficulty that many physicists and philosophers have found in accepting indeterminism in the atomic domain should not blind us to the fact that for Bohr the indeterminacy of the predictions from the formalism was the theoretical reflection of a fundamental fact about the behavior of atoms expressed by the quantum postulate. This was the "surprise" that the investigation of atomic phenomena revealed, thus dictating the need for a new framework. But that surprise had appeared in 1913; thus Heisenberg's discovery of the uncertainty principle in 1927 was no surprise for Bohr, whose whole orientation had prepared him for just such a revolution. For this reason there is relatively little discussion of the statistical character of quantum theoretical predictions in Bohr's work.

Before turning to Bohr's analysis of the uncertainty principle, two points should be made clear. First we should recognize that the mathematical formulas expressing the relations encompassed under the title of the "uncertainty principle" are straightforward deductive consequences of the quantum theoretical representation of the state of the system. Unfortunately discussions of the principle often begin by analyzing a series of thought-experiments intended to demonstrate that observations which determine the value of one observable within a certain range require physical conditions which preclude determining its canonically conjugate observable within a range that would contradict the appropriate uncertainty relation. From presentations of this sort it is easy to get the mistaken impression that the principle expresses an empirical generalization derived from analyzing physical situations and that such a presumed empirical discovery is then injected as a postulate into the theoretical formalism. However, this is not the case. Heisenberg first developed the mathematical formalism for a theoretical representation of quantum mechanical processes, and then showed that the consequence of this formalism was that the system could not be characterized by a state which was defined in terms of precise values of both canonically conjugate parameters. Since our analysis of complementarity does not require studying the mathematical formalism, there is no need to consider the derivation of the principle itself. Our concern, as was Heisenberg's and Bohr's, is to give this theoretical conclusion some physical significance.

In the previous chapter we saw that Heisenberg was led to his discovery by proposing that "nature allowed only experimental situations to occur which could be described within the framework of the formalism of quantum mechanics".[41] In considering various "experimental situations" he, and later Bohr, did indeed try to show that it was physically impossible to determine with arbitrary precision the values of both of the canonically conjugate observables. However, the reason for this impossibility remains notoriously ambiguous in Heisenberg's own presentation.

In some cases his analysis of observation in specific physical situations appears to suggest that the uncertainty arises because the observation "disturbs" the state of the observed system in an uncontrollable way. In this case the uncertainty would seem to be a consequence of the fact that observation requires interaction, and in the interactions described by quantum mechanics, systems change state discontinuously. This "disturbance" interpretation must *accept* the classical presupposition that all physical systems exist in well-defined states theoretically represented by the concept of the classical mechanical state. But an observational interaction disturbs that state in an uncontrollable way so as to make *knowing* the relevant parameters impossible. In this case it must be meaningful to talk about the classical state of quantum systems, even though we cannot know precisely the parameters which would define any particular state, for otherwise we could not say that the observation had disturbed the system from that state.

However, in other places, Heisenberg seems to adopt the more "verificationist" view that since it is impossible to verify empirically statements about a precisely determined classical mechanical state for atomic systems, assertions making use of this concept are *meaningless*. Thus he comments:

> Any use of the words "position" and "velocity" with an accuracy exceeding that given by [the uncertainty relations] is just as meaningless as the use of words whose sense is not defined. [To this sentence he appends the following note:] ... one should remember that the human language permits the construction of sentences which do not involve any consequences and which therefore have no content at all—in spite of the fact that these sentences produce some kind of picture in our imagination. ... One should be especially careful in using the words "reality", "actually", etc., since these words very often lead to statements of the type just mentioned.[42]

One could hardly ask for a clearer statement of the "verificationist" approach to the uncertainty principle. According to this view, associated with the logical positivists who exerted considerable influence at the time, the impossibility of making statements about the atomic system existing in a well-defined classical state lies in the fact that it is meaningless to refer to properties of physical systems considered as anything other than certain observable phenomena. However, before concluding that Heisenberg understood the principle from such a positivistic viewpoint, it is worth noting that in a rather obvious inconsistency, in the same work from which the above passage was quoted, Heisenberg seems to make a claim about what the electron is "really" like: "This indeterminateness is to be considered an essential characteristic of the electron, and not as evidence of the inapplicability of the wave picture."[43] Thus one looks in vain for a single interpretation of the uncertainty principle in the writings of its discoverer.

If we reflect for a moment on Bohr's characteristic approach from an analysis of the applicability of concepts to the description of nature, we can see how differently he approached this principle. Typically the two parameters involved in the uncertainty principle are those of position and momentum or of energy and time. Regarding Heisenberg's discovery from the viewpoint of the complementarity of the two classical modes of description, Bohr saw the determination of position or time as the goal of the mode of space–time coordination whereas the determination of momentum or energy as the goal of applying the "claim of causality" through the conservation principles. In the mathematical formalism used to represent the quantum mechanical state of the system, the degree of determinateness with which the position and time of a system can be predicted is reciprocally related to the determinateness with which its energy or momentum can be predicted. This fact immediately confirmed Bohr's analysis that the goals of the two classical modes of description could not be simultaneously attained. But just as the uncertainty principle allows one to determine position and time *or* momentum and energy with any arbitrary degree of precision, so the two classical modes could both be applied but only in a complementary way.

Thus from Bohr's viewpoint, the uncertainty principle does *not* express any consequence of "disturbing" the system by observation. But his analysis does depend on recognizing that observation involves an interaction which is "uncontrollable" once the quantum postulate is accepted. Instead of assuming that the classical concept of the state of the system does indeed represent the system but that the uncertainty principle shows this cannot be *known*, as does the "disturbance" interpretation, Bohr tried to show that this classical concept of the system's state is an attainable goal only by assuming that it is possible to apply both spatio-temporal and causal modes of description to the system at the same time. Since the quantum postulate entails that this is precisely what cannot be done, the concept of the classical mechanical state of the system is no longer well-defined in application to the objects of quantum mechanical description. For this reason, the disturbance interpretation is totally at odds with the sort of analysis of the uncertainty principle that complementarity advocates. In fact, as we saw in the previous chapter, Bohr had already arrived at the complementarity of the two modes of description before he knew of the uncertainty principle, but he quite naturally saw Heisenberg's discovery as confirming his analysis. We know that for a time he even considered using the word "reciprocal" (derived from the mathematical way of expressing the uncertainty relations between the canonically conjugate parameters) instead of "complementary" to refer to the relationship between the two modes of description.[44]

Though it forms no part of complementarity the "disturbance" interpretation was frequently defended as part of the so-called "Copenhagen Interpretation", and often identified as Bohr's view, in the years immediately following Heisenberg's discovery. From the perspective of Heisenberg's thought-experiments, it appeared that the basis of the principle lies in the fact that instruments of observation "disturb" the observed system such that its state after the observation is no longer that determined in the measurement. This interpretation compares observation of atomic systems to measuring, for example, the workings of a wristwatch with a yardstick. On this view the limitation expressed by the principle results from the fact that our knowledge depends on observations using instruments of a magnitude so grossly larger than atoms that observations intended to determine the state of atomic systems always interfere with that state such that no precise determination of it is possible. This understanding of the uncertainty principle presupposes that interpretation of the classical framework which claims that systems are in some well-defined classical state at all times; the measurement interaction merely alters that state.

However, the disturbance interpretation plays havoc with the actual facts of both the historical genesis of the uncertainty principle and its logical status within the mathematical apparatus of quantum theory. As noted above, the principle is a straightforward deductive consequence of the quantum theoretical formalism which provides a highly confirmed means for predicting the outcome of interactions between radiation and matter. There is no mention of disturbing the system in the derivation of the principle itself, nor of how one goes about determining the relevant parameters. The design of experiments is relevant only in interpreting the physical significance of the principle. Here the assumption that the observed system really exists in a classical mechanical state poses the question of whether an experiment can be designed which will yield greater knowledge about the state of the atomic system than the uncertainty relations permit. If this could be done, nature would allow knowing more than the theory is able to predict, and the theory would be properly judged incomplete.

The mistake of the disturbance interpretation becomes apparent the moment we note that according to it, we could approach the classical ideal of strict determinism if only our instruments of observation were approximately the same size as atoms. However, in fact it is only the immense *difference* between the dimensions of ordinary human experience and those involved in atomic processes that ever made strict determinism a nearly attainable ideal. Were we of the same dimensions as atoms, then the role of the quantum in an interaction would be ever more *increased* rather than decreased, as the disturbance interpretation maintains.

In classical mechanics an observation also "disturbs" the observed system, but the nature and extent of this disturbance is either negligible or "controllable", and so can be accounted for in defining the state of the isolated system after the observational interaction. In the quantum theoretical representation of an interaction ordinarily the effect of the interaction cannot be considered negligible nor can it be controlled. Because the disturbance interpretation makes it appear that the uncertainty principle is an empirical generalization supported by analyses of experimental situations rather than a logical consequence of the formalism, it is unable to explain *why* this alleged disturbance cannot be determined within the quantum framework, and so allow a return to deterministic predictions. Indeed, if one *presupposes* that atomic systems do exist in well-defined classical mechanical states, as the disturbance interpretation must, then one must conclude quantum theory is but a half-way house which in principle could be replaced by a more adequate theory allowing a return to strict determinism. According to such an outlook if a scientific description of nature is possible, then an adequate theory *must* be able to account for (what is regarded as) the "disturbance" involved in observation. The fact that quantum theory cannot do so, on this view, entails that it is an inadequate or "incomplete" theory.

A second possible interpretation, the "epistemic interpretation", holds that the uncertainty principle is the consequence of the fact that the theoretical formalism of quantum theory is developed to describe not the properties of atomic systems but rather what we can *know* about nature as we experience it in observed phenomena. If one accepts the view that the adequacy of a theory is measured only by its success in predicting observable phenomena, then the uncertainty principle gives no reason to hold that quantum theory is inadequate. An analysis of experimental situations reveals that the physical conditions necessary for a position or time observation preclude those which would make a momentum or energy observation possible. Thus the theory is in no way incomplete in not permitting a prediction of simultaneous position and momentum parameters, because no phenomenon ever allows us to determine such information. The fact that the uncertainty principle is a deductive consequence of the theoretical formalism merely expresses this fact and warrants accepting this mathematical instrument for predicting the relevant phenomena.

The epistemic interpretation may be motivated by the viewpoint that talk about the properties of a transphenomenal nature is meaningless and irrelevant to science. While such an outlook may be argued on other grounds, it cannot be inferred from quantum theory alone. Classically the notion of a reality existing apart from its phenomenal manifestations could be regarded as that to which the concept of the classical mechanical state of an isolated system referred. Since such classical mechanical states described the motions of particles or waves through space, the reality behind the phenomena we experience could be consistently regarded as composed of such particles and waves. What quantum theory tells us is that if its description of atomic systems is considered complete, since it cannot define a classical mechanical state for these systems, they cannot be considered "particles" or "waves" in the classical sense. All that follows from this fact is that the notion of a reality independent of phenomena cannot be described as it was previously thought it could be described from the classical viewpoint. It does not necessarily follow that such a concept is without meaning or refers to that about which knowledge is impossible. While the concept of an independent reality can no longer be assumed to be describable as that which possesses properties corresponding to the classical mechanical state parameters, it does not follow that this concept of an independent reality bears no relation to the world as experienced or as described by scientific theory.

If the epistemic interpretation cannot *establish* that the notion of an independent reality is meaningless, we may ask how does it justify its claim that our knowledge is limited in the way expressed by the uncertainty principle? Here the defender of this view is confronted with a dilemma. On the one hand, if he denies the meaningfulness of any claims concerning the character of an independent reality, then he cannot say that the reason we cannot know, for example, the simultaneous position and momentum of a particle is because *in reality* there are no such properties to be known. But if he cannot give any such reason for the limitations of the uncertainty principle, then why not conclude that an altogether different theoretical formalism which describes atomic systems in terms other than those employed by quantum theory might not in fact escape these limitations and allow a return to strict determinism? This seems to have been the expectation of Bertrand Russell as expressed in a letter to Bohr: "I hope to get to understand why Heisenberg's principle is so incompatible with determinism. Left to myself I should only have inferred that the things to be determined are not what used to be supposed."[45] Consequently, if a defender of the epistemic interpretation adopts the view that reference to a reality behind the phenomena is meaningless, he must conclude that it is possible that the limitations of the uncertainty principle may yet be overcome, or in other words that in principle our knowledge is *not* limited in the way the uncertainty relations say it is.

On the other hand, if the defender of the epistemic interpretation does allow the possibility of describing an independent reality, then if the theory is complete in providing everything that can be known about such objects, the fact that there are limits on how precisely we can determine the state parameters must be a consequence of a deeper limitation on what there is to be known. In other words, the limitation is not just a limit on knowledge but reflects a fact about the nature of reality. If this is the case, the basis of the principle is not "epistemic" but "ontological".

Thus either the epistemic interpretation cannot be made consistent with Bohr's claim that quantum theory is complete or it must admit that the limitation on knowledge expressed by the uncertainty principle results from the character of a reality behind the phenomena. It may well be that the limitations of the uncertainty principle do not prohibit the formalism from describing any phenomena *now known* to us. But in the future, experiments might be designed to produce phenomena allowing us to determine more than the current formalism permits. On this epistemic interpretation the fact that nature has thus far hidden from us any means of experimentally determining the simultaneous position and momentum of a particle may reveal only a limitation on our imagination in conceiving experiments. Thus the very best that a defender of the epistemic interpretation could hold is that we may rest content with a formalism restricted by the uncertainty limitations because such a formalism is adequate for the description of all phenomena now known. However such an attitude may well dampen the search for new phenomena that could reveal the incompleteness of the current formalism. The fact that the epistemic interpretation does not base what it claims to be a limitation on knowledge in an essential feature of reality, allows one the possibility to maintain consistently merely that the formalism is incomplete rather than that the description of nature in the atomic domain is incompatible with determinism. If he is barred from any assertions concerning what nature is really like, the defender of the epistemic view can only "explain" the limitation expressed by the uncertainty principle by saying that the theoretical formalism requires it, and, at present we have no better formalism. If we accept the classical view that theoretical parameters do correspond to the properties of an independent reality, then the fact that quantum theory does use such parameters implies that nature possesses the corresponding properties. If this is the case, then the inability of the formalism to be able to define simultaneous values for both canonically conjugate parameters means only so much the worse for the formalism. The proper conclusion then seems to be that the present theory is incomplete and we should search for a more complete one.

The epistemic interpretation of the uncertainty principle, or something like it, was what Einstein seemed to believe Bohr was defending in complementarity. By this time Einstein had totally repudiated the view that science is concerned only with developing a formal means for predicting phenomena. For this reason he repeatedly attempted to devise thought-experiments which would reveal more about the system than the quantum formalism would permit, thereby putting the lie to Bohr's belief that it was complete. Had Bohr in fact defended the epistemic interpretation of the uncertainty principle, as Einstein seemed to believe, Einstein's attitude would express the open-mindedness characteristic of scientific progress and Bohr's would have been dogmatic.

Finally let us turn to Bohr's interpretation of the uncertainty principle. As Bohr understands it, the uncertainty principle reflects theoretically the physical fact that given the quantum postulate, the conditions necessary for unambiguously defining the classical mechanical state of the system as isolated from interaction preclude the physical conditions necessary for giving empirical content to the descriptive concepts in terms of which the state of that system must be defined. He summarizes his understanding of the uncertainty principle at the end of Section 3 of the Como paper:

> According to the quantum theory, just the impossibility of neglecting the interaction with the agency of measurement means that every observation introduces a new uncontrollable element. Indeed, it follows from the above considerations that the measurement of the positional co-ordinates of a particle is accompanied not only by a finite change in the dynamical variables, but also the fixation of its position means a complete rupture in the causal description of its dynamical behaviour, while the determination of its momentum always implies a gap in the knowledge of its spatial propagation. Just this situation brings out most strikingly the complementary character of the description of atomic phenomena which appears as an inevitable consequence of the contrast between the quantum postulate and the distinction between

> object and agency of measurement, inherent in our very idea of observation.[46]

As we saw in the previous chapter, in his debates with Heisenberg, Bohr would insist that in order for the theoretical representation of the physical system to have any empirical content whatsoever, it must be possible to derive from it a description of the observed phenomena in which they are treated as the causal result of the behavior of atomic systems. To give the classical descriptive terms the empirical reference they must have if the theory is to be confirmed by observation requires interacting with the system. But the quantum postulate entails that the interacting whole has an individuality which prohibits unambiguously defining the state of the system while it is being observed. Thus the physical conditions necessary for observation (interaction) are complementary to those necessary for defining the state of the system (isolation). Bohr regarded the uncertainty principle as directly expressing the formal consequence of this complementarity between the mode of space–time co-ordination and the mode of description of causality.

Quantum theory is empirically confirmed by phenomena which are interpreted as physical interactions between the observing instruments and atomic systems. In representing these interactions, the observing system is described in terms referring to the observable properties of the phenomena, *i.e.*, in terms of the "classical concepts". Insofar as the formalism does allow predictions of phenomena described in terms of these classical parameters, these parameters correspond to properties that are just as truly properties of the object being described as they were classically. But Bohr does not take this fact as implying that these properties also belong to an object as it exists apart from such observational interactions. Instead, complementarity holds that the classical concepts refer to properties that belong only to a *phenomenal object*, the object as it is observed.

Since the quantum postulate denies the classical justification for regarding the parameters used to define the state of an isolated system as a picture of the properties of an independent reality, Bohr repudiates the classical correlation of the state parameters with the properties of the object regarded as isolated from any observational interaction. The "pictures" we can form using these concepts of an isolated object refer not to a concrete reality lying behind the phenomena, but to what Bohr calls "abstractions".

> On the whole, it would scarcely seem justifiable, in the case of the interaction problem, to demand a visualization by means of space–time pictures. In fact all our knowledge concerning the internal properties of

> atoms is derived from experiments on their radiation and collision reactions [*i.e.*, on interactions], such that the interpretation of experimental facts ultimately depends on the abstractions of radiation in free space and free material particles [*i.e.*, on systems as isolated]. Hence, our whole space–time view of physical phenomena, as well as the definition of energy and momentum depends ultimately on these abstractions. In judging the application of these auxiliary ideas, we should only demand inner consistency, in which connection special regard has to be paid to the possibilities of definition and observation.[47]

Thus the use of the classical terms to describe atomic systems as "particles" or "waves" interacting with the measuring apparatus refers not to a concrete reality but to an abstraction which is necessary for describing the phenomena as interactions between the measuring agencies and the physical systems described by quantum theory.

The *disturbance* interpretation holds that the objects of the atomic domain really do have properties corresponding to the parameters that would define their classical mechanical state, but the uncertainty principle tells us that the act of measurement is such that it is impossible for us to ever know simultaneous precise values of both parameters. In attempting to avoid all reference to a transphenomenal object, the *epistemic* interpretation must contend that the uncertainty principle is merely a consequence of the formalism we have for describing the phenomena we actually observe. *Bohr's* interpretation is that the limit expressed by the principle is a limit on the applicability of concepts to the description of phenomena. But this limitation is not just the consequence of our concepts, for the limit on the applicability of the concepts is itself the consequence of a fact of nature expressed in the quantum postulate.

Because we must still describe observations in the concepts of the classical framework, from the perspective of that framework we tend to view the uncertainty principle as a *limit on knowledge*, a barrier to ever knowing something which is there and which we would like to know, but never can. However, this misunderstanding of the uncertainty principle is a consequence of presupposing that the descriptive concepts of "particle" and "wave" picture the object to which they refer as it exists in isolation from any observation. If, for example, we regard the electron as really a particle in the classical sense, then, by definition, it has the properties which would define its classical mechanical state at every instant in its trajectory from one point to another. In other words, we fail to make the quantum postulate a true *postulate* in our description of nature. Consequently we imagine that somehow the limitations expressed in the postulate, formally reflected in the uncertainty principle, are not inherent in nature. Thus they can be circumvented, and we can gain empirical information which the formalism precludes. Bohr explains the apparent limitation of the uncertainty principle, by arguing against this classical tendency to interpret the descriptive concepts which have well-defined empirical reference in application to *phenomenal* objects as also referring unambiguously to the properties of an *independently real object*.

Since wave–particle dualism tells us that the concepts through which we represent the state of the system as isolated from interaction cannot refer to the properties of an independent physical reality, Bohr concludes that the concepts by which we express what we have learned in an observational interaction refer to properties of phenomenal objects which one can describe only in the interaction of the system with the observing instruments. The fact that observations of atomic systems can be described as interactions in which the atomic system is represented as a particle in some situations and as a wave in others is a consequence of the fact that these different representations refer to *different* observational interactions each with properties that describe only that interaction.

By recognizing the complementarity between space–time co-ordination and causal description, Bohr avoids the dilemma that gives trouble to the epistemic interpretation. Since each phenomenal appearance is a unique individual interaction, descriptions referring to *different* observational interactions occurring in different experimental situations must be treated as observations of *different phenomenal objects*. If one holds the view that the concept of the object existing apart from its phenomenal manifestations is meaningless, then it becomes impossible to say that descriptions of different phenomena are *complementary* in the sense that together they exhaust all it is possible to say about the *same* system, for there is no "same system" in the sense of the phenomenal objects observed. Thus it would always be possible to hold that a different set of concepts might be able to describe these different phenomena as all the consequences of interactions in which the object system with which the observing instruments interact is described in some unitary way, not as a particle in one instance and as a wave in another. To speak of wave–particle dualism would be merely to make a comment about this particular way of representing the phenomena which are treated as the consequence of atomic behavior. Wave–particle dualism would then not be a feature of reality, but a feature of our way of representing it, and the possibility would be open for a different description which would remove such a dualism. The fact that Bohr emphatically believed that such a search was *futile* reveals that he regards the complementarity of different phenomenal descriptions to be the result of a fact of nature as we observe it, and as is expressed by the quantum postulate.

The complementarity of space–time descriptions and causal descriptions is also a result of the concepts we employ. But Bohr's claim is that we have *no choice* in using these classical concepts, for they are the ones which refer unambiguously to the properties of phenomena as observed. Thus when he says that the limitations of the uncertainty principle are a consequence of the concepts we use to describe observation, he does not mean to suggest that other concepts could be developed to represent these phenomena in a way which would allow us to know more about the behavior of atomic systems than Heisenberg's principle permits.

The classical causal space–time description of observational interactions made it possible to define the state of the system as isolated from any interaction while observing it. The theoretical representation of the state of such systems as particles or waves in "free space" isolated from any interaction imparts to the classical framework the descriptive ideal of picturing the system as it exists apart from observations. In holding that the uncertainty principle is a consequence of the nature of atomic systems, and for this reason wave and particle pictures of a system isolated from observation are but abstractions, complementarity violates the classical descriptive ideal which co-ordinated theoretical parameters with properties of an independent reality. For complementarity, the fact that the formalism of quantum theory has found it necessary to accept the quantum postulate demonstrates the need for restructuring the relationship between theoretical parameters and the world they are used to represent. As Bohr would say, we must relearn the presuppositions governing the use of our most elementary concepts. Thus complementarity tells us that since the parameters of a physical theory have *empirical* significance only if interpreted as referring to specific properties of observed phenomena, when those same parameters are used to define the state of the system isolated from observation, we cannot assume that they can be interpreted as referring to properties possessed by that system apart from observational interactions. At the same time, this new framework cannot be regarded as coherent if it ignores the question of how to refer to this "independent reality" which in interacting with observational instruments produces the phenomena which are described through the theory. Facing up to the epistemological and ontological consequences of this fact will concern us in the last two chapters, but before we turn to these problems, it will be useful to see how Bohr refined his doctrine in the light of Einstein's criticisms and how he extended it to fields beyond atomic physics.

CHAPTER FIVE

THE REFINEMENT OF COMPLEMENTARITY

As noted previously, Bohr's Como paper made little impact on its audience; it was not a paper in physics, and presented no new empirically testable consequences of quantum theory. The prevailing mood of dissatisfaction among atomic physicists was so great that they expected a resolution to the quantum mystery in a much more concrete and clear-cut manner. Nevertheless, the Como paper did make it clear that Bohr now regarded the development of the quantum revolution as in a certain sense "complete". Quantum theory need no longer await some enlightening revelation which would put everything right from a classical point of view. This stance at first surprised, puzzled, and left unimpressed a great many of Bohr's colleagues.

Bohr intended his Como presentation as a general outline for a new framework to be worked out in greater detail later; thus in the years following 1927, his attention turned to refining the arguments of complementarity. Naturally the application of complementarity to quantum physics was always foremost in Bohr's mind, but as early as 1929 it is obvious that he had a strong interest in carrying its epistemological lesson into other fields. Curiously for the history of complementarity, his thought did not develop as he originally seems to have believed it would into a generalization of the classical framework extending over all empirical knowledge. Instead its course was sharply influenced by the criticisms of Einstein. It is of course impossible to say just how different Bohr's philosophical work would have been had he received strong endorsement from this crucial source. Nevertheless, it is clearly the case that a great deal of Bohr's thought, and in a sense his finest victories, were manifested in specific reply to challenges put forward by Einstein.

Although the differences between Bohr and Einstein reflected a very real disagreement about the nature of the scientific description of physical reality, as has been suggested, there were also serious misunderstandings on both sides. It must be admitted that Bohr probably weakened his own cause by not stressing from the beginning that ultimately what was at stake was no less than a proposed revision of our understanding of the relationship between physical reality and the concepts used to describe it. In a certain sense Einstein realized this more clearly than Bohr himself, but he would have nothing of it. For his part, Bohr was very sensitive to the effect of the quantum postulate on the use of concepts, but he appears to have remained somewhat indifferent to the repercussions his analysis of the use of concepts in describing nature would have on the scientists' conception of physical reality.

1. The Bohr–Einstein Debates on the Nature of Physical Reality

The debate over complementarity which lasted the remaining thirty years of Bohr's life did not formally begin until one month after the Como conference, when Bohr and Einstein met at the Solvay Congress in Brussels. Here Einstein opposed Bohr's views by appeal to the first in a series of thought-experiments focusing on finding a way to get around Heisenberg's uncertainty relations so that the classical notion of the state of the system could be retained. In 1927 Bohr had little difficulty in showing how Einstein's first thought-experiments failed to provide more empirical information about the system than permitted by the quantum mechanical formalism, but at the Solvay Congress of 1930 Einstein presented a particularly difficult thought-experiment. After a sleepless night, Bohr's memorable response involved demonstrating how Einstein had failed to consider the effects of his own general theory of relativity, which when taken into account gave precisely the degree of uncertainty to the measurement that was compatible with Heisenberg's principle.

In 1935, in conjunction with B. Podolsky and N. Rosen, Einstein presented his final attempt at a *Gedanken-Experiment* designed to show that the physical system had simultaneous properties that quantum theory could not determine, thereby demonstrating that the theory was "incomplete". This experiment, commonly known as the "EPR experiment", finally brought into the open the opposition between two conceptions of physical reality. Indeed Bohr concluded his reply to the EPR challenge with the claim that "complementarity" is a "new feature of natural philosophy [which] means a radical revision of our attitude as regards physical reality ...".[1] However, the nature of the difference between complementarity and the classical framework is such that, at the physical level, there cannot be an "*experimentum crucis*" demonstrating the correctness of one view over the other. Thus from the point of view of complementarity, Bohr's answer is completely consistent, while from the point of view of the classical framework Einstein's objections are equally consistent.[2] Of course Einstein's rejection of the completeness of quantum theory allowed him to adhere to the classical framework, while Bohr's acceptance of its completeness forced him to adopt a new framework.

For our purpose of understanding complementarity as a conceptual framework for describing nature, it is not necessary to analyze in detail the theoretical description of Einstein's thought-experiments; Bohr himself has provided an excellent non-technical review of the debates which hardly could be improved upon.[3] However, we do need to recognize the fact that many of Bohr's writings after 1927 were composed with the implicit goal of winning over Einstein or his partisans. This fact has had both beneficial and detrimental consequences. On the one hand, it enabled Bohr to review and to refine the application of his viewpoint to specific phenomena, both in discussion and in print. But on the other hand, it changed his original focus from a general argument concerning "the description of nature" and the use of a "conceptual framework" in that description to a detailed analysis of the physical situation in numerous imaginary experiments involved with observing atomic systems. Without discounting the importance of the experimental foundations of Bohr's application of complementarity to quantum theory, it may be said fairly that such a method was hardly conducive to understanding the general philosophical themes that Bohr originally had hoped to develop. Since he was urging an entirely new conceptual framework for describing nature, attempting to "refute" his claims by analyzing the experimental situation from the *classical* viewpoint produced little more than a yet sharper division of opinion between the adherents of complementarity and its opponents.

In this manner complementarity as a general philosophy of natural science gave way to the "Copenhagen Interpretation" of quantum theory, and Bohr's concern with the fundamental revision of the framework of scientific description was largely lost from sight. Nevertheless, the never settled debate between Einstein and Bohr was merely the first round in the still persisting controversy over the proper interpretation of quantum theory. By following that debate only to the extent to which it gave Bohr's interests a specific direction, this chapter will simply try to understand that development rather than try to solve the problems which divided the two views.

In his last interview, Bohr was asked this question: "Einstein was always apparently asking for a definition or a clarification, a precise formulation of what is the principle of complementarity. Could you give that?" Bohr replied, perhaps bluntly, "He also got it, but he did not like it."[4] This simple reply may indicate the magnitude of the missed communication between these two scientific giants. As the question suggests, Einstein seemed to read complementarity as a *principle* in the physical sense and was thus frustrated by Bohr's failure (in his eyes) to provide a clear formulation which would have testable consequences in the fashion of physical principles in science, parallel in this respect to Einstein's own principle of relativity. Failing to find such a principle, Einstein interpreted Bohr's position as denying the view that the aim of science is to describe the nature of an independently existing physical reality.

Einstein argued from the assumption that properties corresponding to the classical state parameters belong to physical entities independently of observations, an assumption which he considered justified because such parameters have empirical reference to observed phenomenal properties which are used to confirm theory. From the fact that in the classical framework it is always possible in principle to define precisely the mechanical state of the system even as it is being observed, it follows that if the parameters which define the state are interpreted as corresponding to properties possessed by the system whose state they are used to define, these systems must in reality always simultaneously have such properties as a precise position and momentum. Therefore, the inability of quantum mechanics to define such states for systems in the atomic domain implies either that there is a physical condition which makes knowing these states impossible or that the states are in principle knowable, but quantum theory is "incomplete" in failing to provide the full measure of potentially possible knowledge about them.

Since Bohr clearly insisted that quantum theory *is* complete, Einstein concluded on his assumptions that Bohr opposed the view that the task of scientific theory is to describe the nature of physical reality. In his private correspondence Einstein gives clear evidence of this understanding of complementarity, when in reference to his arguments against Bohr's view, he speaks of hoping to "overcome ... the abandonment of the concept of reality in physics".[5]

Perhaps unwisely, Bohr tried to win over Einstein by comparing his own arguments concerning quantum theory with the sort of arguments that the young Einstein had used in developing relativity. This fact may well have strengthened Einstein's conviction that Bohr was approaching the quantum problem from the point of view that all a theory need do is predict correctly the observable phenomena. When Philipp Frank, who was indeed a partisan of this view, tried to convince Einstein by pointing out the similarity of Bohr's approach to that which Einstein had himself used, the latter replied, "A good joke should not be repeated too often."[6] Hence Einstein presumed complementarity to be based on a rejection of the view that the task of science is to describe the properties of an independent physical reality, a position which the elder Einstein's whole scientific orientation found repugnant.

Thus Einstein himself, and many other observers, believed that the debate with Bohr required defending the view that it is the intention of scientific theory is to describe the properties of physical reality as it exists independently of our observations. The debate was never brought to the level of recognizing the possibility of *different* conceptions of such a description of "physical reality" against which the question of a complete or incomplete description of this reality is to be set. Since the prevailing mood in philosophy of science during this period was sharply influenced by the positivistic distrust of discussion of "the nature of physical reality", it was not in the interest of such philosophers of science to present the debate in this way. Only in the last decades has the true significance of this historic interchange slowly come to be realized.[7]

2. The Einstein–Podolsky–Rosen Experiment

Einstein and his colleagues eventually abandoned the attempt to disprove quantum mechanics by finding a way to make *simultaneous* observational determinations of parameters which would defy the uncertainty relations. Instead they focused their efforts on demonstrating the incompleteness of quantum theory by considering an arrangement in which it is claimed that one can use observation and theory to infer that the system has simultaneously both a precise momentum and position.

To argue that a theory is incomplete, it is necessary to stipulate what constitutes a complete theory. As a criterion of completeness, Einstein suggested: "every element of physical reality must have a counterpart in physical theory".[8] Clearly such a criterion becomes meaningful only when we can specify what constitutes "physical reality". Here, Einstein provides the following sufficient condition: "If without in any way disturbing a system, we can predict with certainty (*i.e.*, with probability equal to unity) the value of a physical quantity, then there exists an element of physical reality corresponding to this physical quantity."[9]

In the EPR experiment the authors consider a situation in which two physical systems, the initial states of which are known, are allowed to interact for a finite period of time. As we have noted, in the quantum formalism once the systems interact, it becomes theoretically impossible to define the states of each system separately. Nevertheless, given the theoretical representation of their initial states as defined in the quantum formalism, it is possible to define the state of the two systems combined, treated in this sense as an individual interacting whole. What cannot be defined by the theory is the state of each system considered separately either during the interaction between the two or after that interaction. Thus the only way the parameters defining the state of either system can be determined after the interaction is by another observation. But to make an observation, the observing system must interact with the observed in a way such that the conditions required for determining, for example, the position of a system will preclude the conditions necessary for simultaneously determining its momentum. Thus, as the uncertainty principle demands, *observation* cannot determine both parameters necessary to define precisely the state of either of the systems after their interaction. However, from the observation of the position of the first system and the theoretical definition of the combined states of the two systems, one is able to determine *theoretically*, *i.e.*, to "predict", the position of the second system without in any way interacting with that second system. Similarly a momentum determination made by observing the first system will allow a theoretical prediction of the momentum of the second system. Bohr does not dispute this conclusion, for he admits that "it is always possible in quantum theory, just as in classical theory, to predict the value of any variable involved in the description of a mechanical system from measurements performed on other systems, which have only temporarily been in interaction with the system under investigation".[10] Since we have a "free choice" whether to set up an experiment observationally determining the momentum or position on the first system, it would seem that, given the assumption that "position" and "momentum" refer to properties possessed by an independent reality, it is possible to determine simultaneously the precise values of two observables *both* of which the quantum theory cannot predict. Thus, since we can know more about the system than quantum theory allows, we must conclude that the theory is incomplete.

Of course determining the momentum of the first system by observation will require an interaction with it that will alter its state so that determining its position simultaneously would be impossible. If the two systems were still interacting, then a change of the state of the first system would be expected to alter the state of the second system. But what if the two systems have long since ceased to interact? If we then determined by observation, say, the momentum of the first system and the position of the second, the knowledge of the momentum of the first system together with the theoretical definition of the state of the two interacting systems would allow one to predict the *momentum* of the *second* system. It would appear that no matter how the state of the second system is changed by the observational interaction which determines its *position*, since by premise the two systems are no longer interacting, there could be no effect on its *momentum* caused by the observation of the momentum of the *first* system. Since in principle we have a "free choice" in choosing to measure either the position or the momentum of the second system, it appears that we could have confirmed the prediction of its momentum by an observation, although of course then we would not have been able to determine by observation its position. If this is the case, then by the "reality criterion", we must conclude that simultaneous reality be ascribed to both the position and the momentum of the second system. The inability of the quantum formalism to represent the system in a way which would permit assigning precise values to both of the parameters corresponding to these properties then shows that this formalism is incomplete in the sense of not allowing knowledge of all that can be physically determined concerning the system.

This conclusion was precisely what Einstein hoped to draw. Nevertheless, the authors do admit that ...

> one would not arrive at our conclusion if one insisted that two or more physical quantities can be regarded as simultaneous elements of reality *only when they can be measured or predicted.* On this point of view, since either one or the other, but not both simultaneously, of the quantities [momentum] and [position] can be predicted, they are not simultaneously real. This makes the reality of [position] and [momentum] depend on the process of measurement carried out on the first system, which does not disturb the second in any way. No reasonable definition of reality could be expected to permit this.[11]

Indeed, on the assumptions that "position" and "momentum" refer to properties possessed by the systems independently of observation and that what the observing system records in an observational interaction is the causal effects of such properties, the claim that measuring the value of a phenomenal observable which is caused by the property of one system alters the value of a phenomenal observable which is caused by the property of another system with which it is in no way physically interacting would seem to demand a mysterious non-physical communication between the two systems. Such a claim about physical reality does indeed seem unreasonable as EPR claim.

However, if we accept Bohr's view that those same theoretical parameters do *not* refer to the properties of an independent reality but instead refer only to phenomenal properties, then if there is no interaction to observe, say, the position of the system, there is no phenomenal property to which the position parameter can refer. In this case, Einstein's conclusion seems equally unreasonable. Thus although we have a "free choice" of which experiment to perform, that freedom merely indicates a freedom in which phenomenon we choose to bring about:

> ... in the phenomena concerned we are not dealing with an incomplete description characterized by the arbitrary picking out of different elements of physical reality at the cost of sacrificing other such elements, but with a rational discrimination between essentially different experimental arrangements and procedures which are suited either for an unambiguous use of the idea of space location, or for a legitimate application of the conservation theorem of momentum. Any remaining appearance of arbitrariness concerns merely our freedom of handling the measuring instruments, characteristic of the very idea of experiment. In fact, the renunciation in each experimental arrangement of the one or the other of two aspects of the description of physical phenomena—the combination of which characterizes the method of classical physics, and which therefore in this sense may be considered *complementary* to one another—depends essentially on the impossibility, in the field of quantum theory, of accurately controlling the reaction of the object on the measuring instruments, *i.e.*, the transfer of momentum in the case of position measurements and the displacement in case of momentum measurements. ...
>
> ... we are, in the "freedom of choice" offered by the ... [EPR] arrangement, just concerned with a *discrimination between different experimental procedures which allow of the unambiguous use of complementary classical concepts.*[12]

If we determine by an observational interaction, say, the position of the second system, then no observational determination of its simultaneous momentum is possible. Thus even though the observation of the momentum of the first system enables one to calculate theoretically the momentum of the second, since there can be no interaction in which that momentum could be determined (because the position observation physically precludes it), that *phenomenal* property to which such a theoretical prediction refers does not exist. Such a theoretical calculation refers to merely an "abstraction".

Of course the theoretical prediction can be confirmed by an observational interaction, as is necessary for such predictions to have empirical significance, but then the interaction which determines the position of the second system is physically precluded. If these theoretical parameters refer only to the properties of observed phenomenal objects, if there is no observational interaction which determines the position of the second system, there is no phenomenal object possessing the property to which the term "position" refers. These alternative interpretations of the EPR experiment reveal the heart of the difference between two distinct conceptions of physical reality.

Using his idiosyncratic vocabulary, Bohr expresses his reply to Einstein by trying to show that Einstein's notion of "physical reality" is "ambiguous". Thus Bohr summarizes his reply as follows:

> ... we now see that the wording of the above-mentioned cirterion of physical reality proposed by Einstein, Podolsky, and Rosen contains an ambiguity as regards the meaning of the expression "without in any way disturbing the system". Of course there is in a case like that just considered no question of a mechanical disturbance of the system under investigation during the last critical stage of the measuring procedure. But even at this stage there is essentially the question of *an influence on the very conditions which define the possible types of predictions regarding the future behavior of the system.* Since these conditions constitute an inherent element of the description of any phenomenon to which the term "physical reality" can be properly attached, we see that the argumentation of the above-mentioned authors does not justify their conclusion that the quantum-mechanical description is incomplete.[13]

The "ambiguity" on which Bohr focuses his analysis was hidden in Einstein's use of "the state of the system". In one sense, when we determine the state of the system by observation, the "system" to which we refer is necessarily the *phenomenal* object. However, in another sense, when the observable properties of the phenomenal object are considered the effects of properties possessed by an independent physical reality, the "system" whose state is determined is presumed to be the object existing independently from observation. In a paper written in 1939, four years after his reply to the EPR paper, Bohr argues ...

> the very fact that in quantum phenomena no sharp separation can be made between an independent behavior of the objects and their interaction with the measurement instruments, lends indeed to any such phenomenon a novel feature of individuality which evades all attempts

> at analysis on classical lines, because every imaginable experimental arrangement aiming at a subdivision of the phenomenon will be incompatible with its appearance and give rise, within the latitude indicated by the uncertainty relations, to other phenomena of similar individual character.[14]

Bohr agrees that one may predict both position and momentum and that one has a free choice on which parameter to measure in confirmation of the theoretical prediction. However, if we choose one alternative, say position, we do so by means of an experimental arrangement which precludes the physical conditions necessary to determine momentum. Of course the reverse holds as well. Bohr does not dispute the "reality" of the physical system as it is observed in an interaction. His point is that the quantum postulate implies that we cannot define, *i.e.*, give empirical reference to, the classical concepts used to define the "state" of a system separately from the state of the observing system with which it is interacting. In 1939 he explained his point as follows:

> In fact the [EPR] paradox finds its complete solution within the frame of the quantum mechanical formalism, according to which no well-defined use of the concept of "state" can be made as referring to the object separate from the body with which it has been in contact, until the external conditions involved in the definition of this concept are unambiguously fixed by a further suitable control of the auxiliary body.[15]

Thus these descriptive concepts are well-defined only in reference to *observed phenomenal* objects. As we shall see, Bohr also does not dispute the *reality* of the system considered as an object existing independently of observational interactions. Indeed, he does argue that the theoretical representation of "the state of the isolated physical system", in this sense an "independent reality", refers to an abstraction, but his reason for this claim is not that no such independently real object exists or that the concept of such an object is "meaningless". Instead, his point is that there can be no empirical warrant for assuming that the classical state parameters refer to the properties of such an entity, and if one does make such an assumption, quantum theory becomes involved in the contradiction of asserting that in the atomic domain systems have the incompatible properties of both particles and waves.

From Einstein's point of view it seems that we have a "free choice" of which of the two possible measurements to make on *one and the same system*. Nevertheless, if we make Bohr's assumption that in order to give terms like "position" and "momentum" empirical significance, "system" must be interpreted in the sense of that which is observed, then since the experimental arrangements necessary for the alternative observations of position and momentum are mutually exclusive, we must conclude that the two observations refer to the properties of *two distinct phenomenal objects*. Thus on Bohr's view a position measurement must describe a different phenomenal object from that which is described in a momentum determination. There is simply no "same system", in the sense of a phenomenal object whose position *and* momentum one could have observed.

In short, the argument of EPR assumes that in the interaction necessary to determine the value of a parameter representing a phenomenal observable, it remains possible to describe the system independently of the observational interaction as having properties corresponding to those which characterize its phenomenal appearances. Thus if the observing apparatus is changed, such that a momentum observational interaction instead of a position observational interaction takes place, Einstein assumed that it does not influence the "property" of the "system", for the property, being the property of an independently real object, has "reality" independently of its phenomenal manifestations.

Bohr was in fact disheartened by the nature of Einstein's reasoning, for it missed the core of his argument from the quantum postulate as presented in Como. Regarding the quantum postulate as axiomatic, Bohr believed Einstein dogmatically refused to accept its logical consequence that we must abandon the classical presupposition that the classical state of the system refers to properties possessed by an independent physical reality. In fact Einstein was far from regarding the quantum "postulate" as axiomatic. For him the quantum mechanical description of the discontinuous change of the atomic system from one stationary state to another must be controlled by as yet undiscovered physical laws which he expected would eventually be discovered and allow a return to the classical ideal of describing change of state continuously. Since all of Bohr's argumentation was erected on the quantum postulate, it all bypassed Einstein. Professor Jammer summarizes the difference between the two views as follows:

> To understand correctly the views of the two disputants it should be recalled that for Bohr these thought-experiments were not the reason but the necessary consequence of a much more profound truth [*i.e.*, that expressed by the quantum postulate] underlying the quantum mechanical description and, in particular, the uncertainty relations. Bohr consequently had the advantage that, from his point of view, he was justified in extending the chain of reasoning until he could appropriate-

> ly resort to the indeterminacy relations to support his thesis. Einstein, on the other hand, had the advantage that if he could disprove the Heisenberg relations by a closer analysis of the mechanics of one single thought-experiment, Bohr's contention of the incompatibility of a simultaneous causal and space–time description of phenomena and with it his whole theory would be refuted.[16]

Recounting their discussions, Bohr quotes Einstein's attitude: "To believe this [*i.e.*, the completeness of quantum mechanics] is logically possible, but it is so very contrary to my scientific instinct that I cannot forgo the search for a more complete conception." To which Bohr's rejoinder confirms this analysis: "Even if such an attitude might seem well balanced in itself, it nevertheless implies a rejection of the whole argumentation ... aiming to show that in quantum mechanics we are not dealing with an arbitrary renunciation of a more detailed analysis of atomic phenomena, but with a recognition that such an analysis is *in principle* excluded.[17] Given Bohr's route to complementarity as we have followed it, since he always took the quantum postulate as his point of departure, it is not surprising that he found it difficult to accept that Einstein did not see this postulate as unavoidable. But Einstein's commitment to the view of reality derived from the classical description was sufficiently strong that he could only read the quantum as introducing a discontinuity into describing physical processes at the cost of ignoring a more fundamental set of laws which would describe atomic transitions in a continuous fashion.[18] In spite of the fact that he had contributed so essentially to this description, to the last, Einstein considered his own use of the quantum as but a heuristic device, ultimately to be discarded. In Bohr's words, "he tried to get the quantum out of the world".[19] Of course, from Bohr's perspective the discovery of the physical fact expressed by the quantum postulate was hardly to be regretted. Not only did it provide the physical key to uniting an atomic conception of matter with continuous field theories of radiation, but it also precipitated an advance in our understanding of how concepts are used in describing nature.

The missed communication between Bohr and Einstein is so momentous that an additional simple but telling analysis of its roots is excusable. As we have noted, since Einstein was by no means inclined to regard the quantum postulate as axiomatic, he could not agree with Bohr's argumentation designed to show from the quantum postulate that the state of an isolated system could no longer refer to an independently real object. Consequently Einstein tended to regard what Bohr was saying as the *apparently* similar (but in fact very different) claim that the concept of the unobserved system as an independent reality was *meaningless*, because by its very definition this concept can refer to nothing which is even in principle observable. Such a position would be supportable only by holding that statements which cannot in principle be empirically verified are meaningless statements. The historical fact that at the time the dominant philosophy of science, logical positivism, did espouse such a "verifiability criterion of meaning" contributed to this misunderstanding of Bohr's position. But we can distinguish Bohr's view from that of positivism simply by noting that on the one hand the positivist bars discussion of what cannot in principle be observed solely on the basis of a logical analysis of scientific methodology, an analysis which may well be mistaken, but which cannot be refuted by any *empirical* discovery of a particular science. On the other hand Bohr bases his argument on precisely such an empirical discovery of atomic physics: that expressed in the quantum postulate. Had there never occurred such a need to resort to discontinuity, *this* conflict between Bohr and Einstein never would have occurred.

3. Bohr's Concept of "Phenomenon"

After his reply to the EPR challenge, Bohr altered his argumentation by emphasizing that the description of nature is free from ambiguity and metaphysical dogmatism only if it is realized that the observational basis of science depends on devising a way to describe unambiguously an individual observational interaction between physical systems. Since this view was already implicit in the Como paper, it is misleading to represent these changes as a modification of Bohr's viewpoint. However, there were very real changes in manner of speaking and in emphasis.

Following 1935 Bohr revised his argument from that presented in the Como paper by distinguishing his view from that of the "disturbance" interpretation, for there he did indeed speak of observation as "disturbing" the system. To speak in this way implies that "system" refers to an independent reality, the state of which is altered by an observational interaction. Of course this way of speaking is entirely consistent with the classical framework and is just how one would expect to describe an observation in that framework. As we have seen, in his reply to EPR Bohr claimed the ambiguity involved in not distinguishing the system as an observed phenomenal object from the system as an independent reality led to the invalidity of Einstein's argument. So it is obvious that after 1935 he must be careful to distinguish the observed phenomenal object from any independently existing physical object.

In our analysis of the Como paper in the preceding chapter, we observed that though much of Bohr's language might lend itself to the "disturbance" interpretation, this view was almost certainly not what Bohr intended to defend. Nevertheless, throughout the essays of *Atomic Theory and the Description of Nature* Bohr does speak in ways that suggest that the object of observation really does have the properties that would define its classical mechanical state, but the observing interaction disturbs that state in a way which makes knowledge of them impossible. In particular, if he refers to the observed atomic object by the term "particle", the suggestion is made that it is a physical object possessing both position and momentum. This way of speaking persisted as late as 1932:

> In fact, the unambiguous application of such fundamental concepts as space and time is essentially limited on account of the finite interaction between the object and the measuring tools, which, as a consequence of the existence of the elementary quantum, is involved in any measurement. To appreciate this point, we must remember that this interaction cannot be taken fully into account in the description of the phenomena, since the very definition of the space–time frame implies the neglect of the reaction of the object on the measuring instruments. Thus, any attempt to fix the space–time co-ordinates of the *constituent particles of an atom* would ultimately involve an essentially uncontrollable exchange of energy and momentum with the measuring rods and clocks which prevents an unambiguous correlation of the dynamical behavior of the atomic *particles* before the observation with their later behavior. Inversely, every application of conservation theorems, for instance to the energy balance in atomic reactions, involves an essential renunciation as regards the pursuance in space and time of the individual atomic particles.[20]

Here Bohr's language easily lends itself to supporting the view that quantum theory forces us to "renounce" the possible knowledge of properties of particles, even though it is meaningful to talk about the constituents of atoms as "particles" and so presumably to assume that such particles really have properties that would define their classical state.

However, in his first philosophical writings after EPR, in 1937, Bohr is careful to point out that there is no question of an "arbitrary renunciation" of the measurement of the properties which really are possessed by the objects which are observed in the phenomena that confirm quantum theory. What we must "renounce" is the classical attempt to "visualize" the objects of quantum mechanical description as possessing "such inherent attributes as the idealizations of classical physics [*i.e.*, "waves" and "particles"] would ascribe to the object":

> ... in these fields the logical correlations can only be won by a far-reaching renunciation of the usual demands of visualization. It would in particular not be out of place in this connection to warn against a misunderstanding likely to arise when one tries to express the content of Heisenberg's well-known indeterminacy relations ... by such a statement as: "the position and momentum of a particle cannot be simultaneously measured with arbitrary accuracy". According to such a formulation it would appear as though we had to do with some arbitrary renunciation of the measurement of either the one or the other of the two well-defined attributes of the object, which would not preclude the possibility of a future theory taking both attributes into account on the lines of classical physics. From the above considerations it should be clear that the whole situation in atomic physics deprives of all meaning such inherent attributes as the idealizations of classical physics [*i.e.*, "particle" and "wave" pictures] would ascribe to the object. On the contrary, the proper role of the indeterminacy relations consists in assuring quantitatively the logical compatibility of apparently contradictory laws which appear when we use two different experimental arrangements, of which only one permits an unambiguous use of the concept of position, while the other permits the application of the concept of momentum defined as it is, solely by the law of conservation.[21]

Thus Bohr makes it obvious that such classical terms as "position" and "momentum" are "deprived of all meaning" apart from the context of their application to describe particular observational interactions of phenomenal objects as they appear in specific observational interactions. However, we should note that his point is that the classical descriptive terms are without meaning when applied to the concept of an independent reality as visualized through the classical "idealizations" of waves and particles. He does *not* assert that the very notion of such an independent reality is itself without meaning or that there is no need to refer to such an atomic object in the description of observation as interaction. As we shall see, this becomes a crucial point in understanding the philosophical significance of Bohr's new viewpoint.

In the Como paper Bohr spoke of "disturbing the *phenomenon*" where "phenomenon" clearly meant what I shall henceforth call the "phenomenal object", the object whose properties the observation determines. In this usage, the "phenomena" are what interact with the observing instruments to produce observations; this usage is conspicuous throughout *Atomic Theory and the Description of Nature*, where he tells us that ...

> ... the fundamental postulate of the indivisibility of the quantum of action is itself, from the classical point of view, an irrational element which inevitably requires us to forgo a causal mode of description and which, because of the coupling between phenomena and their observation, forces us to adopt a new mode of description designated as *complementary* in the sense that any given application of classical concepts precludes the simultaneous use of other classical concepts which in a different connection are equally necessary for the elucidation of the phenomena. ...
>
> ... *the finite magnitude of the quantum of action prevents altogether a sharp distinction being made between a phenomenon and the agency by which it is observed*, a distinction which underlies the customary concept of observation and, therefore, forms the basis of the classical ideas of motion.[22]

Even as late as 1937 Bohr spoke of complementary "aspects of quantum phenomena revealed by experience obtained in mutually exclusive conditions", implying that the same phenomenon could appear with different aspects in different observational interactions.[23] However, this way of speaking made it seem that one could unarbitrarily distinguish between the observing system and the "phenomenon" which is observed in an interaction. Since his reply to Einstein emphatically disavowed this way of describing observation, from 1939 onwards Bohr altered his use of the word "phenomenon" to refer now to *the whole observational interaction*, the description of which forms the empirical basis for confirming the theory:

> The unaccustomed features of the situation with which we are confronted in quantum theory necessitates the greatest caution as regards all questions of terminology. Speaking, as is often done, of disturbing the phenomenon by observation, or even of creating physical attributes to objects by measuring processes, is, in fact, liable to be confusing ["misunderstanding" in the original draft] since all such sentences imply a departure from basic conventions of language which, even though it sometimes may be practical for the sake of brevity, can never be unambiguous. It is certainly more in accordance with the structure and interpretation of the quantum mechanical symbolism, as well as with elementary epistemological principles, to reserve the word "phenomenon" for the comprehension of the effects observed under given experimental conditions.
>
> These conditions, which include the account of the properties and manipulation of all measuring instruments essentially concerned, con-

> stitute in fact the only basis for the definition of the concepts by which the phenomenon is described.[24]

This stipulation makes it clear that Bohr regards the different experimental arrangements for determining space–time coordination or determining the exchange of momentum or energy (and so requiring different interactions with the atomic system) as "*different* types of quantum phenomena". Thus in the same article where he introduces this terminology, he explains the complementarity of modes of description using this definition of "phenomena" as follows:

> The apparent contrast between different types of quantum phenomena, the description of which involves different classical ideas like space–time coordination or momentum and energy conservation, finds in fact its straight-forward explanation in the mutually exclusive character of the different experimental arrangements demanded for the appearance of such phenomena. Thus, any phenomenon in which we are concerned with tracing a displacement of some atomic object in space and time necessitates the establishment of several coincidences between the object and the rigidly connected bodies and movable devices which, in serving as scales and clocks respectively, define the space–time frame of reference to which the phenomenon in question is referred. Just this situation implies, however, a renunciation of any sharp control of the amount of the momentum or energy exchanged during each coincidence between the object and the separate bodies entering into the experimental arrangement. Inversely, every phenomenon in which we are essentially concerned with momentum and energy exchanges—and which therefore necessitates an experimental arrangement allowing at least two successive determinations of momentum and energy quantities—will, in principle, imply a renunciation of the control of any precise space–time co-ordination of the objects in the time intervals between these measurements.[25]

In succeeding articles he repeatedly called attention to this terminology, noting that the "rational account" of a "phenomenon ... must be taken to involve a complete description of the experimental arrangement as well as the observed results".[26]

Adopting the word "phenomenon" to refer to "the effects observed under given experimental conditions" had a significant impact on Bohr's expression of complementarity after 1939. As we saw in the previous chapter, Bohr's leading idea in formulating complementarity centered on the complementarity of two modes of description, that of space–time co-ordination and that of the claim of causality. However, because the debate with Einstein showed the tendency to regard position observations and momentum observations as determinations of the properties of the *same* observed system (*i.e.*, the *same* "phenomenon" as Bohr used that term in the Como paper), Bohr began to emphasize that these two observational interactions are *different* phenomena and hence determine the properties of *different phenomenal objects*. Thus he eventually adopted a way of speaking which referred to the complementarity of different *phenomena* or complementary *evidence* from different observations.[27] For this reason some physicists have argued that Bohr's doctrine underwent a radical change and/or that there are really two distinct notions of complementarity.[28] Although Bohr would agree that complementarity is a many faceted viewpoint, in fact when confronted with these two different uses of "complementary", Bohr resolutely maintained that the complementarity of modes of description and the complementarity of phenomena are simply two consequences of the quantum postulate.

The appearance that there are two distinct doctrines of complementarity is abetted by Bohr's shift in his use of the word "phenomenon". As early as 1927 in the manuscripts for the Como paper, he says, "the wave and corpuscular ideas are able only to account for complementary sides of the phenomena".[29] In the published paper he altered his choice of words and speaks instead of the complementarity of "wave and particle pictures".[30] Nevertheless, this way of speaking suggested that *one phenomenon* could have two complementary "aspects", that described by the particle picture and that described by the wave picture, and as we have seen above, Bohr did in fact speak this way in 1937.[31] Even as late as 1939, after introducing his new use of the word "phenomenon" in the passage cited above, Bohr continued with the comment that "phenomena defined by different concepts, corresponding to mutually exclusive experimental arrangements, can be unambiguously regarded as complementary aspects of the whole obtainable evidence concerning the objects under investigation".[32] Hence when he refers again to "complementary phenomena" after 1939, one might assume he is referring to "wave phenomena" and "particle phenomena" in the sense of two distinct "aspects" of entities in the atomic domain as we observe them. Such a view seems quite distinct from the position that complementarity exists between space–time co-ordination and causal description. Since a phenomenon which permits being described as an observational interaction determining position or time excludes a phenomenon which permits being described as an interaction determining energy or momentum, these phenomena are "complementary" as a consequence of the complementarity relationship existing between the mode of space–time description and the mode of causal description to determine momentum or energy parameters through the conservation laws. It is true that in one of these phenomena we may describe the observed object as a "particle" and in the other as a "wave". This fact thus warrants Bohr's talk of the complementarity between wave and particle pictures, but this complementarity is, so to speak, "derived" from the former. It should not mislead us into assuming that the "particle picture" is to be identified with descriptions of phenomena which determine space or time parameters or the "wave picture" with phenomena determining momentum or energy.

The first indication of a changed way of describing the complementary relationship appears in 1938, where the relationship is held to occur between pieces of "information":

> Information regarding the behavior of an atomic object obtained under definite experimental conditions may . . . be adequately charcaterized as *complementary* to any information about the same object obtained by some other experimental arrangement excluding the fulfillment of the first conditions.[33]

But after the war, beginning in 1946, he begins to deemphasize the complementarity between space–time description and applying the conservation principles in favor of defining the complementary relationship as holding between *phenomena*: "Although the phenomena in quantum physics can no longer be combined in the customary manner, they can be said to be complementary in the sense that only together do they exhaust the evidence regarding the objects, which is unambiguously definable."[34] This sentence, or some variant of it, becomes a standard formula for Bohr's later work and is repeated over and over again in almost every essay. However, he does continue to use other formulations as well, for example:

> . . . we are faced with the contrast revealed by the comparison between observations regarding an atomic object, obtained by means of different experimental arrangements. Such empirical evidence exhibits a novel relationship, which has no analogue in classical physics and which may conveniently be termed "complementarity" in order to stress that in the contrasting phenomena we have to do with equally essential aspects of all well-defined knowledge about the objects.[35]

As we shall see in the last chapter, the fact that this way of speaking indicates that complementary phenomena provide different pieces of evidence about the *same* atomic object implies that the atomic object cannot be identified with the different complementary *phenomenal* objects which are described through the classical wave and particle pictures.

Therefore, once it is recognized that Bohr's talk of complementary phenomena after 1939 refers to the effects observed under different experimental conditions, then the temptation to regard such talk as creating a confusion between complementary descriptive modes and complementary aspects of nature disappears. In fact all of Bohr's analyses of experiments in the debates with Einstein were intended to show that determining the spatio-temporal properties of the observed object is possible only in experimental interactions which physically preclude any interaction determining momentum or energy properties of the object. Thus to give empirical reference to their terms, spatio-temporal co-ordination and causal description require physical interactions, "phenomena" in Bohr's latter usage, which exclude each other but are complementary. Since the experiments involving different observational interactions between observing apparatus and object system are what Bohr calls "phenomena" after 1935, the fact that certain "phenomena" are complementary is a logical consequence of the complementarity of space–time and causal descriptions. Thus an experimental arrangement which produces a "phenomenon" that is described as an interaction determining position physically excludes any arrangement which would produce the different "phenomenon" that is described as an interaction determining momentum. The complementarity between a space–time description and a description applying the claim of causality to determine momentum or energy implies the complementarity of two distinct phenomena. Since these complementary phenomena are treated as observations determining the values of different parameters, we can speak of the "complementary evidence" or the "complementary information" provided by these different phenomena. In this way Bohr can justify his rejection of the interpretation that there are two distinct notions of complementarity.

4. The Reformulation of the Object of Description

The debate with Einstein caused Bohr to realize that there was a fundamental ambiguity hidden in the notion of "physical reality" as that which is described in natural science. Thus he was moved by these discussions to modify his presentation of the complementaristic analysis of the object of description.

Physical science bases its theories on the descriptions of experimental observations. Thus at the empirical end of description, the prediction from theoretical representation to experimental results, the object to which the descriptive concepts must refer is the *phenomenal* object. Insofar as science describes certain phenomena as "observations" which determine the properties of systems in the atomic domain, it must treat the indivisible phenomenon as an interaction between systems, thereby forcing a distinction between the observing system and the observed phenomenal object. To do so, in the description of the phenomenon as an observational interaction we must draw a "partition" between the interacting physical systems, although in the quantum theoretical representation of the interaction we cannot define separate states for these interacting systems. Bohr emphasizes that the "individuality" of the whole observational interaction makes any attempt to "subdivide" the phenomenon an *arbitrary* distinction imposed for the sake of *describing* the phenomenon of interaction as an observation of some phenomenal object. But in order for observation to determine unambiguously the values of phenomenal observables, the description of the observation must indicate exactly what part of the whole interaction is considered observing instruments and what part is considered phenomenal object being observed. For this reason an inherent arbitrariness is introduced into the description of any phenomenal object, for the description of the interaction may have partitioned the whole phenomenon differently, making what is now part of the observing system part of the observed or *vice versa*. Nevertheless, only by making such an arbitrary separation between what is treated as the means of observation and what the phenomenal object can an unambiguous description of *what* is observed be communicated. This is the reason for the "ambiguity" of Einstein's attempt to determine the state of what *must* be a *phenomenal* object, without referring to the interaction in which the observation takes place.

Thus Bohr speaks of drawing a necessary but arbitrary distinction between observing and observed systems. This distinction is "necessary" because in order to describe unambiguously an interaction as an observation of a particular object, the distinction between it and the agency of observation must be stipulated. But where that distinction is made is "arbitrary", in the sense that there is no theoretical way to define the classical mechanical state of the systems thus distinguished. "Observing system" and "observed object" are terms which are well defined only in the context of a particular *description* of an interaction. Hence they must be regarded as *descriptive* categories invoked for an unambiguous communication of the results of an observation rather than as referring to different constituents of nature. Although the distinction between observing system and observed object is arbitrary from the physical point of view, the context in which the description of a phenomenon is to be employed in science effectively determines the particular distinction which is made. Thus for the purposes of measurements which would serve to confirm quantum theoretical predictions, the whole individual phenomenon will be described in a way which distinguishes between what is treated as an atomic system, or constituent thereof, and the instruments which are described as determining a particular property attributed to that system.

In a certain sense Bohr had already made these points in the Como paper. Thus as early as 1927 (well before Einstein's criticism), we find in his manuscripts for Como the comment that the fact that observing atomic phenomena involves "an essential interaction ... means not only that through the observation the phenomena are essentially influenced, but strictly speaking our only means of defining the concept involves interaction".[36] But the debate with Einstein focused his attention on the fact that in an observation the terms we use to describe the phenomenal object refer to an "object" which is defined only in the context of a particular description of a whole individual phenomenon. Thus the debate caused him to realize ever more fully that the central point of his conceptual revolution involved at its heart the "epistemological lesson" that the fundamental problem in the philosophical account of science is understanding the relationship between descriptive concepts and the objects they are used to describe.

Although the quantum postulate expresses the crux of the revolution from which Bohr originally derives the force of his argument, after Como he progressively deemphasized the derivation of complementarity from the quantum postulate, stressing instead the physical basis in the individuality of the phenomenon for the applicability of descriptive concepts. Nevertheless, over and over again he points out that the assumptions on which the classical framework was based have proved strictly false. Thus in the later essays, although he does not present his case as derived from the quantum postulate *per se*, he makes it clear that complementarity is the result of the physical fact which the quantum postulate expresses, namely the individuality of the phenomenon which is described as an observational interaction. He saw this physical fact as expressing a feature of "atomicity" on which the atomic constitution of matter was based. Thus he says that this indivisibility of interaction in the quantum formalism reveals that there is "a feature of atomicity in the laws of nature going far beyond the old doctrine of the limited divisibility of matter [which] has indeed taught us that the classical theories of physics are idealizations which can be unambiguously applied only in the limit where all interactions are large compared with the quantum".[37] Within the atomic domain the individuality of an interaction requires that in using the classical concepts we must revise our understanding of their descriptive function. Bohr clearly held that the way we can use concepts in describing nature is determined by which physical assumptions prove acceptable in the light of the empirical confirmation of theories making use of these assumptions. If the assumption that physical systems can be described as existing in classical mechanical states even while interacting with other systems turns out to be empirically unsupported, then in using the classical concepts to report the results of an observation, the use of those concepts to define the classical state of the system must be understood in a new light. Thus we must relearn the use of the classical concepts, or in other words, we must understand the sort of objects to which they may unambiguously refer.

We have noted that the phenomenal object described in an observation is distinct from the object about which we have complementary evidence or information. Since the *phenomenal* object is defined only in the context of a particular observational interaction, and since interactions which permit applying the mode of space–time description exclude interactions which enable one to determine momentum or energy using the conservation principles, it follows that the *phenomenal* objects described in these two complementary modes are *different* objects. Thus if they provide complementary information about the *same* object, as Bohr claims they do, the "object" in this latter sense cannot be the phenomenal object. One is thus led to infer that the "atomic object" is the object which appears in different ways in these distinct phenomenal interactions. This conclusion, however, *may* seem incompatible with Bohr's insistence that the use of the classical descriptive terms to picture the state of an isolated system refers to an abstraction. Thus we must be careful to recognize that Bohr does *not* mean to say that the concept of anything but the phenomenal object is meaningless.

Bohr does *not* argue that since giving a term some empirical content ("defining" it in his words) requires it to refer to a phenomenal object, it must follow that reference to unobservable entities is "meaningless". Bohr holds that the reason the classical descriptive terms (which must have empirical reference so that they can be used to describe phenomena) cannot be used to describe an "independent reality" is a result of a *physical fact*, namely the indivisibility of an observational interaction as implied by the quantum postulate. The requirement that descriptive terms must have some empirical reference if they are to describe phenomenal objects makes their attempted use to describe an atomic object apart from its phenomenal appearances in observational interactions descriptively "ambiguous": "... the impossibility of subdividing the individual quantum effects and of separating the behavior of objects from their interaction with the measuring instruments serving to define the conditions under which the phenomena appear implies an ambiguity in assigning *conventional attributes to atomic objects* which calls for a reconsideration of our attitude towards the problem of physical explanation."[38] However, we should note at once that this passage indicates that Bohr does not hold that the concept of the "atomic objects" (as that about which the complementary phenomena provide all empirically determinable information) is itself without precise meaning. What he tells us quite plainly is that the "conventional attributes" (*i.e.*, those which define the classical state of the physical system) cannot be unambiguously predicated of such atomic objects. Bohr never argues that the notion of the object as an independent reality is *meaningless*. What he does argue is that because of a certain physical fact, the individuality of atomic interactions ("a feature of atomicity in the laws of nature going far beyond the old doctrine of the limited divisibility of matter"), the classical descriptive concepts in terms of which particle and wave pictures are defined become restricted in their reference to the phenomenal object. Therefore the use of *these* concepts to *picture* the state of an isolated system does not refer to an independent reality but to an "abstraction". Bohr's point is that we cannot describe an independent reality in the manner which was thought to be possible in the classical framework, because that framework made different presuppositions about the nature of observation.

An example involving a somewhat more familiar concept than those of quantum physics, the concept of temperature, may help clarify Bohr's position. Considered as a descriptive concept applied to large sized bodies of many molecules, temperature can be unambiguously defined by giving it empirical reference through the specific operations by which temperature is determined. If in the theoretical representation of the properties of material bodies it had been empirically confirmed that describing such bodies as continuously divisible was an adequate description, then there would have been no conceptual barrier against expecting that "temperature", like "mass" or "position" would refer to each bit of matter no matter how far it was subdivided. However, it was discovered empirically that material bodies can be adequately described by representing them as composed of many molecules and that the property of temperature observed in many-molecule bodies can be represented as the causal effect of the motion of these molecules. For this reason, these physical facts make the concept of temperature inapplicable below the molecular level.

It may seem a fine distinction to make between the philosophical position that talk about an independent reality is "meaningless" and Bohr's claim that the classical concepts are "inapplicable" to such a notion because the physical conditions for their definition require interaction. Indeed if one identifies meaning with "verifiability", an empirically inapplicable descriptive concept would be "meaningless", since there would be no way to test assertions using it. However, Bohr's point depends on an empirical discovery, whereas that of the defender of a verifiability criterion of meaning does not. Thus "temperature" could have been applicable to all levels of material bodies no matter how far subdivided, had the structure of matter turned out to be adequately represented as continuously divisible according to the best of our theories. It is simply a contingent fact that we are now led to accept an atomic theory which represents matter as not so divisible. For one who accepts a verifiability criterion of meaning by *fiat*, statements referring to the properties of objects which by definition are unobservable (as is an independent reality), become meaningless as a logical consequence of the definition of meaning quite independently of any empirical discovery. But Bohr's concern lies with the applicability of these concepts for describing *phenomena*. It must be a matter for empirical investigation to determine whether or not the physical conditions necessary for giving such concepts empirical reference are actually the case. The lesson of complementarity is that the extension of scientific knowledge into new areas, in this case penetrating the structure of matter below the molecular level, can produce surprises and can become instructive about the presuppositions governing the proper use of such concepts.

According to Bohr's position, the classical notion of the state of a physical system would have been applicable to quantum mechanical objects if the classical presupposition of the continuous change of state had produced a theory which met with predictive success as knowledge was extended into the atomic domain. If this was the case, the quantum postulate would never have been accepted and interactions in the atomic domain would not have the individuality which Bohr claims they must have. Had this logical possibility occurred, then the descriptive ideal of determining the classical state of an atomic system while it is observed would have been well-defined, and hence an attainable goal. Thus the classical justification for regarding the theoretical representation of the isolated system as a picture of the properties of an independent reality would have been upheld. In such a world, Einstein's criticism, based as it was on rejecting the quantum postulate, would have hit its mark. But since the classical presupposition on which this representation rested has been empirically disconfirmed, any use of the spatio-temporal concepts to define the state of the system as isolated from observation is understood to refer to an abstraction. Thus in Bohr's epistemological lesson we learn that at the atomic level, as a consequence of an empirical fact, the classical spatio-temporal concepts are not defined in an unambiguous sense for an objective description of an independent physical reality.

CHAPTER SIX

THE EXTENSION OF COMPLEMENTARITY

During World War II work on the philosophical problems in physics more or less ceased; when Bohr returned to writing on complementarity after the war, the position of atomic physics had changed radically. There now existed a whole profession of atomic physicists, who had attained a degree of specialization beyond the reach of the older men for whom the intellectual struggle of the quantum revolution had involved central philosophical concerns. Quantum theory could be used to solve countless problems in atomic physics, and further theoretical progress did not seem to require clarification of the troublesome philosophical issues raised by the quantum revolution. As Jammer observes, "Impressed by the spectacular successes of quantum mechanics in all fields of microphysics ... [the vast majority of physicists] were interested primarily in its applications to practical problems and in its extensions to unexplored regions. ... The need of acquiring new mathematical techniques left little room for philosophical analysis."[1]

Bohr continued to campaign for complementarity on a broad basis, but his fame as an atomic physicist in a sense stood in his way, for it was virtually only in application to quantum theory that anyone investigated complementarity at all, and even then only in the restricted application of dealing with complementary "wave pictures" and "particle pictures" in particular experimental situations. Bohr had envisioned complementarity spreading out into wider and wider fields, just as the mechanical approach of Galileo had started in astronomy and simple phenomena of motion and gradually spread to all of the physical sciences. However, thus far, history has proved him unduly optimistic, for instead physicists and philosophers have concentrated on more and more detailed problems and have developed increasingly sophisticated analyses of the theoretical formalism. Undoubtedly Bohr was dismayed and discouraged by these developments. In an effort to counter this tendency and to put complementarity into a more general light, in the remaining chapters discussion of quantum problems as such will be avoided except where necessary to illustrate Bohr's arguments.

Examining Bohr's writings from 1935 to his death in 1962, one cannot help concluding that his overriding philosophical goal was to bring the lesson of complementarity to fields other than atomic physics. His articles from this period are often very repetitious, but that should only serve to deepen our *prima facie* conviction that he did not consider these applications of his epistemological lesson to biology, psychology, and anthropology to be merely instructive parallels. Instead, he was fully convinced that the generalization of the classical framework which he called complementarity would teach the lesson for revising our understanding of the description of nature in these fields as it had in physics.

As noted in the Introduction, most discussions of complementarity restrict themselves to analyzing Bohr's interpretation of quantum theory. While that interpretation certainly deserves careful analysis, such treatments have almost uniformly ignored Bohr's point that complementarity teaches a lesson about the use of concepts for describing all phenomena. Thus complementarity has tended to be of interest only to those concerned with the philosophical foundations of physics. Those few analyses which do deal with complementarity outside of atomic physics have tended to commit the opposite error of not securing any connection between a vaguely defined "principle of complementarity" and its roots in the quantum revolution. In order to right this imbalance, this chapter will begin by considering how we can take our leave of the quantum postulate in atomic physics and generalize the epistemological lesson it has taught us to include fields beyond the quantum domain.

1. Complementarity and Empirical Knowledge

Now that we have traced how Bohr came to develop complementarity, it is not hard to find the reason for this relative scarcity of discussions of complementarity beyond atomic physics. It would appear that Bohr has backed himself into a corner from which he cannot easily extricate himself. We have already observed that he always begins his arguments with the empirical discovery expressed by the quantum postulate. From this fact and the demand for an unambiguous description of phenomena in science, Bohr claims we must inevitably have recourse to a complementary combination of modes of description.

Thus Bohr's argument seems to imply that in those sciences concerned with the description of phenomena that arise in observational interactions where the dimensions of the interaction are such that the specific quantum effects may be considered negligible, one may strive consistently to attain the older classical ideal of simultaneously observing an object and defining how it exists apart from that observation. If this conclusion is true, it would appear that complementarity could teach an epistemological lesson only in those fields where the quantum postulate has detectable consequences, or in other words, where the behavior of atomic systems directly enters the description of the phenomena pertinent to that field. One might, then, expect complementarity to be relevant in a field like molecular biology but hardly expect it to have anything to teach in psychology or anthropology.

However, against this conclusion we can set Bohr's explicit claims concerning the epistemological lesson of complementarity in a variety of fields quite distinct from atomic physics. Although he occasionally ventured comments on the relevance of complementarity to art, music, or religion, in general he is concerned only with empirical knowledge, *i.e.* the sciences. We know from his deep conviction of the "unity of knowledge" that he believed that the quantum revolution would ultimately lead to a general complementaristic philosophy of empirical knowledge. The importance of Bohr's commitment to a single framework for describing all phenomena is indicated by the fact that this phrase, "the unity of knowledge", appears as the title of two of his articles and recurs as a theme in numerous others. The grounds for this "unity" of the sciences lies in the fact that all are concerned with the objective description of phenomena which are observed in human experience. Thus Bohr speaks of empirical sciences as sharing the common "problem of observation", or the problem of devising a framework suitable for the unambiguous description of the pertinent phenomena. Since complementarity teaches how the exploration of new phenomena may force revising the presuppositions concerning the applicability of descriptive concepts, and thereby entails a generalization of the framework within which such descriptions are offered, its lesson may well be relevant to all empirical sciences as they grow to encompass more and more phenomena.[2]

Of course the class of observations where Planck's quantum plays a non-negligible role is but a small subclass of all possible observations used by empirical science. Thus the epistemological lesson taught by the discovery of the quantum postulate can be no more than to make us aware of the fact that classical physics had used its presupposition of the continuity of interaction to justify its view that observation either did not alter the observed object or that the effect of observation could always be taken into account. In this way it was possible to regard the description of the object isolated from observational interaction as a picture of an independent reality even though the observations which provide the empirical grounds for that description are made on an object interacting with the agencies of observation. As is implied in the claim that complementarity is a generalization of the classical framework, that presupposition is only true in a special case and cannot be assumed to be true in general.

Since all science aspires to an unambiguous description of phenomena based on observation, a discovery about the nature of observation potentially affects all sciences. Of course it was recognized long before the quantum revolution that the minimal necessary condition for an observation was an interaction between observed object and the agencies of observation. As we saw in the previous chapter, an unambiguous description of this interaction as an observation requires distinguishing clearly between the observed object and the observing system. In the classical framework, which extended its ideals over a wide region of science, it was considered to be a criterion of objective description that the properties ascribed to the object were possessed by it quite apart from the observational interaction on which the description was based. A description which attributes to its object properties that exist only as relations between the object and the observing system in Bohr's terminology is "ambiguous", because the reference of the descriptive terms in one observation may well change when only one term of that relation, the observer, changes. Since the individuality of the interaction means that the properties of the *phenomenal* object do exist only in relation to the specific agencies of observation which produce that phenomena, the description is unambiguous (*i.e.*, "objective") only when it includes a full description of the agencies of observation which produce the phenomenon being described. One could make the determination of the state of a physically isolated system the criterion of objective description only if such a description could be regarded as picturing the properties of an independent reality. However, as we have seen, such a descriptive ideal is consistent only in the special case where either the causal effect of the observational interaction can be "controlled" or regarded as negligible.

The quantum postulate entails that this is not the case in describing atomic systems. The discontinuity of change of state gives the whole an individuality which makes it impossible to define the state of the observed object after the observation. Thus the effect of the observation cannot be determined more precisely than the uncertainty principle allows. Moreover, in atomic interactions the magnitude of this discontinuity is such that the way in which the observational interaction changes the state of the observed object cannot, in general, be ignored. Thus the physical conditions required for *defining* the state of the isolated system (non-interaction), preclude those necessary for *observing* it (interaction). However, in order to describe an interaction as an observation, it is necessary that one can define theoretically the state of the object as isolated from interaction. So we cannot dispense with this mode of description. At the same time if the description is to be confirmed by empirical means, the object must be observed; hence this mode of description also cannot be abandoned. Bohr's solution is to consider these two distinct descriptive modes as "complementary", each pursuing a goal necessary for the employment of the other. The complementary combination of the two yields an unambiguous description of all that observation can reveal concerning the physical systems described in atomic physics. The price paid for this way of securing an objective description is abandoning the view that the theoretical representation of the state of the isolated system pictures the properties of an independent reality. Once this epistemological lesson is learned, the paradox of wave–particle dualism generated by viewing the description in the classical framework is dispersed.

Just as complementarity taught this lesson to atomic physics, so it may well need to be learned in other fields of empirical science where the effect of an observational interaction can neither be ignored nor precisely determined. Bohr expresses this lesson in his oft-repeated "admonition" that "it must never be forgotten that we ourselves are both actors and spectators in the drama of existence".[3] The effect of learning such a lesson in other empirical sciences may be the need to revise the philosophical understanding of their description of phenomena.

Of course in learning this lesson, other sciences will see the specific descriptive modes of physics (space–time co-ordination and use of the conservation principles to determine the mechanical causal development of such a space–time description) as necessarily limited to the task of physics. Each field, then, would "define the state" of its objects and describe the causal effect of observational interaction on that state in terms of concepts appropriate to that field. The task of each science is to develop a framework of concepts which permits achieving these goals adequately for all known phenomena which fall under its scope. As that scope is extended the framework may need to be suitably "generalized".

Strictly speaking, descriptions expressed in terms of a discontinuity in the change of state can be meaningful only for those sciences which describe their objects in terms of states defined by quantitative parameters which may vary "continuously" or "discontinuously". Hence the extension of complementarity to other fields must generalize from this notion of discontinuity to considering interactions where the change in the observed object can neither be ignored nor predicted deterministically. Although the example of an anthropologist doing field research on a primitive tribe is not how Bohr attempted to apply complementarity to anthropology, it is too tempting in this context to ignore.[4] Obviously in a case where an anthropologist from an "advanced" culture observes the behavior of a tribe hitherto isolated from outside "interactions", his presence may well affect the behavior of the tribe in a way which cannot be be considered negligible nor can it be controlled. Yet if he describes just what he observes, his description will be ambiguous because another observer coming in different circumstances may well alter the tribe's behavior in a different way. Bohr's conclusion would be that neither description describes the "real" tribe as it exists apart from these observations of it. What both describe is the tribe's phenomenal manifestation of its behavior patterns in different specific observational interactions. In order to overcome this ambiguity, the anthropologist must first recognize that descriptions of the phenomena he encounters must include an account of his interaction with the tribe in a way which produces his observations. But in order to combine this description with other complementary phenomena which are produced by the tribe's interaction with other observers, the anthropologist also develops causal principles which enable him to determine how a particular interaction from outside alters a tribe member's behavior. Applying these causal principles, he can then "reconstruct" a theoretical representation of the tribe's behavior when it was isolated from his observing interactions. By applying these causal principles to this theoretical representation of the isolated tribal behavior, the anthropologist can predict the ways in which the tribe will change its behavior as a result of his appearance on the scene. When these predictions are confirmed, he comes to accept his theoretical representation as adequate. But he must recognize that this well-defined representation is but an abstraction necessary to secure unambiguous description, not a "picture" of the "independently existing tribe".

It may be that Bohr never used this example because unfortunately it does not present a perfect analogy to the situation in atomic physics. The tribe and the anthropologist do not form an interacting whole such that any distinction between the two is arbitrary. However, if the anthropologist stays long enough, he may become himself enough of a *de facto* member of the tribe to make arbitrary any distinction between himself as now tribe member and himself as outside observer. In any event Bohr did not argue this case, but he did think that he found in psychology and biology instances of precisely such observational interactions which did partake of a "feature of individuality" that made the observational interaction inherently unpredictable in its causal effect on the state of the observed object.

The key to Bohr's extension of complementarity beyond atomic physics is the fact that in descriptions of certain phenomena the required observational interaction has this indivisible quality. From this situation it follows that the interaction is "uncontrollable", thus forcing us to regard the representation of the object as isolated from observational interaction as an abstraction. This conclusion in turn implies that a proper understanding of science must prohibit regarding the terms in which such an abstraction is expressed as corresponding to the properties of an independent reality. Bohr believed that he had discovered violations of precisely this prohibition in controversies where rival descriptions of some range of phenomena were interpreted as disputes about the nature of independently real objects. Specifically, he found such situations in the "free will/determinist" controversy, which he interpreted as a dispute in psychology, and the "mechanist/vitalist" controversy in biology. In both cases he believed that the resolution of the dispute, like the case of wave–particle dualism, would not involve a victory for one side or the other, but would necessitate "a renewed examination" of "the possibilities of observation and definition":

> I am far from sharing, however, the widespread opinion that the recent development in the field of atomic physics could directly help us in deciding such questions as "mechanism or vitalism" and "free will or causal necessity" in favor of the one or the other alternative. Just the fact that the paradoxes of atomic physics could be solved not by a one sided attitude towards the old problem of "determinism or indeterminism", but only by examining the possibilities of observation and definition, should rather stimulate us to a renewed examination of the position in this respect in the biological and psychological problems at issue.[5]

He also tended to see a similar situation in "nature or nurture" disputes in cultural anthropology; however, this extension of complementarity was never fully developed and was strongly colored by Bohr's desire to counter Nazi racial theories of the time. Here I will consider only the first two controversies on which Bohr spent any significant amount of effort, those in psychology and biology.

2. Complementarity in Psychology

We have seen that as an adolescent Bohr listened in on the discussions between Høffding, the philosopher, and his father, the physiologist, and that later he joined in the discussions of Høffding's students in the Ekliptika Circle. On these occasions his interest in epistemology was aroused by the problem of describing human conscious processes, particularly as this question touched on the freedom of the will. As already discussed briefly in Chapter Two we know from Bohr's last interview that while at the University studying with Høffding, he had the idea to represent the difficulties in describing one's own consciousness through the analogy with Riemann's analysis of multivalued functions of a complex variable. As noted there, his only account of these ideas is far from clear:

> At that time I really thought to write something about philosophy, and that was about this analogy with multivalued functions. I felt that the various problems in psychology—which were called the big philosophical problems, of the free will and such things—that one could really reduce them when one considered how one really went about them, and that was done on the analogy to multivalued functions. ...[6]

Bohr never put these ideas in writing and his explication of them in his last interview is, by his own admission, "very, very obscure". The basic scheme, however, seems to be as follows. In attempting to describe one's own consciousness, *i.e.* the consciousness one directly experiences, one must inevitably make a distinction between that consciousness as the object of description and the subject consciousness which experiences it. This distinction is essentially Kant's distinction between empirical and transcendental egoes. All attempts to describe the experiencing subject in its experiencing activity, the "transcendental ego", necessarily elude the grasp of the descriptive concepts, for as soon as one attempts to describe that subject, it becomes the object of the experience, the "empirical ego", thereby shifting the distinction between experiencing subject and experienced object. If one is not careful to observe this shifting distinction, the attempted discription becomes hopelessly ambiguous, referring to different egoes each time the distinction is implicitly moved. It is easy to see the immediate connection of this line of thought to that expressed humorously by Poul Martin Møller in Bohr's favorite tale, *Adventures of a Danish Student*, already discussed in Chapter Two. Like the Licentiate, the incautious would-be observer of his own conscious states becomes lost in a labyrinth of his own egoes generated by being unaware of the shifting reference, the ambiguity, of his descriptive concepts.

Without presuming advanced knowledge in mathematics, the analogy of this situation to multivalued functions of a complex variable can be given in a simple qualitative way. The theory of complex functions is one area of mathematics where multivalued functions arise. Each complex number can be represented unambiguously as a point on a two-dimensional plane, but the multivalued function may have potentially an infinite number of values, each a complex number, for each value of the independent complex variable. Thus if one were to map a function involving, for example, the logarithm of a complex number in a single plane, then the representation would be ambiguous, each point standing for many different values of the function. Hence Riemann proposed mapping such functions as different "branches" of a single curve, each on a different plane, and each one representing the curve of a single-valued function. In this case, as long as one follows the curve in the same plane, the function can be mapped continuously without any ambiguity arising in the point number relationship. However, for each function the origin point on the plane is a "singular point", resulting in the mathematical consequence that when a closed curve is traced around the origin, such that it returns to the same value of the independent variable, the value of the function now differs by a constant factor. Were we to continue to map the function in the same plane, the value of the function would become ambiguous in the graphical representation. For this reason, each time we orbit a singular point, the value of a function must be represented in another plane.

Now turning to the analogy between this situation and the description of consciousness, Bohr explains the parallel as follows:

> The analogy is this, that you say that the idea of yourself is a singular in our consciousness. ... Then you find—now it is really a formal way—that if you bring this idea [*i.e.*, the idea of the experiencing subject] in, then you leave a definite level of objectivity or subjectivity. For instance, when you have to do with the logarithm, then you can go around; you can change it by 2π when you go one time around a singular point. But then you surely, in order to have it properly and be able to draw conclusions from it, will have to go all the way back again in order to be sure that the point is what you started on. Now I'm saying it a little badly, but I will go on. That is then the general scheme, and I felt so strongly that it was illuminating for the question of free will, because if you go round, you speak about something else, unless you go really back again [the way you came]. That was the general scheme you see.[7]

Though the details are lacking, it is clear that Bohr intended to correlate the relationship between the independent complex variable and the different values of a multivalued function with a descriptive term and the different references the term could have. Riemann had proposed eliminating this mathematical ambiguity of a point on a single plane representing many possible values of a complex function by turning the multivalued function into a series of single valued functions represented in different planes. So Bohr proposed that the different references of a term refer to different "planes of objectivity", each analogized to a different meaning imparted to the "object" of the experience as a consequence of different ways of drawing the distinction between experiencing subject and experienced object. The chance for ambiguity arises if we fail to note that when we trace a closed curve around the singular point of the origin, we must move to a different plane. Bohr intended to draw the lesson that the attempt to describe the self is like drawing a closed curve around a singular point. The concept of the self, then, is analogized to a multivalued function, which may take on different references. Thus in attempting to describe the subject self (transcendental ego), we must "map" that meaning onto one plane of objectivity, but in doing so we make that subject self the object and thus effectively shift the subject/object distinction. Therefore when we return to the subject self, it is not the "same" self as was the subject before we began to describe it. We must then recognize that the reference of the term "self" has moved to another plane of objectivity. A description unheedful of his fact runs the risk of having its terms liable to slip from one plane of objectivity to another, thereby producing a confusing "ambiguity".

Bohr never tells us how he imagined this analogy would solve the problem of free will, but the following is one possible argument he may have intended. When a subject reports willing to do an action freely, he gives an entirely accurate and truthful description of his experience. However, here the awareness of the freedom in willing to do his act is considered part of the *subject's* state in the experience; the *object* of description is the action which he reports willing to perform freely. Therefore in such a description the descriptive term "freely willing" refers to a completely subjective "feeling" accompanying the subject's awareness of choosing to do an act. Hence it must be regarded as belonging to the subject's side of the subject/object distinction, where that distinction stands in the original report, "I freely choose to do this act". We can, of course, shift this distinction and describe this act of freely willing to do a certain action as itself an object, but in doing so the single phrase, "the act of freely willing", including the subject's feelings in doing that act, now refers to what falls wholly on the object side and cannot refer to the (now different) "subject" whose conscious awareness is the basis for describing the whole experience as an experience of willing to do an act freely Thus we pass from one plane of objectivity to another by shifting the subject/object distinction, thereby rendering the description of "the act of willing" systematically ambiguous. If we fail to discover this ambiguity, we think that we are describing the *same* object when, describing the experience of a subject choosing to do a certain act with the feeling of acting freely on the subject's side, we describe the willing as taking place freely, and then when, describing the experience with the subject's feeling of freedom as part of the *object of description*, we claim that this phenomenon, including the feeling of freedom, was itself the product of deterministic laws operating on our previous psychological states. If we heedlessly imagine these descriptions as referring to one and the same "object", we misunderstand the use of concepts in the description of this experience. Hence, we transform the *descriptive* distinction between free will and determinism into a distinction between two rival notions of the nature of the "self", the object of description in psychology, one answering to deterministic laws, and the other not. Just as the same sort of ambiguity in the concept of "physical reality" ("phenomenal object" or "independent reality") created the apparently contradictory claims that the object of quantum mechanical description has the properties of waves and has the properties of particles, so being unaware of the ambiguity in the concept of "consciousness" generates the apparent contradiction that the "self" has the incompatible properties of being both free and determined at the same time with respect to the same act. However, careful analysis of the use of descriptive concepts shows that it is not the case that one description is correct and the other an illusion to be explained away. Instead, they are understood to be complementary descriptions, in each of which the partition between subject and object self has been drawn differently. Thus if we are aware of the ambiguity in the relevant concepts, the problem of free will versus determinism will never arise.

While Bohr may have explored this line of thought well before complementarity appeared, it should be clear that this early approach falls perfectly in line with the epistemological lesson of complementarity. As early as 1927, Bohr had concluded his Como lecture with the rather cryptic remark, "I hope, however, that the idea of complementarity is suited to characterize the situation, which bears a deep-going analogy to the general difficulty in the formation of human ideas, inherent in the distinction between subject and object."[8] Here is evidence that from its inception, Bohr conceived complementarity as teaching an epistemological lesson concerning any description of phenomena where the wholeness of an observational interaction makes a distinction between subject and object arbitrary. Bohr explicitly states this lesson with respect to psychology in the first paper he wrote after Como, for the Planck *Festschrift* in 1929. He begins by referring to the last sentence of the Como paper and then continues with what is probably his most revealing description of the general epistemological situation as he saw it:

> At the conclusion of the [Como paper] ..., it was pointed out that a close connection exists between the failure of our forms of perception, which is founded on the impossibility of a strict separation of phenomena [here before 1937 Bohr means the phenomenal object] and means of observation, and the general capacity to create concepts, which have their roots in our differentiation between subject and object. ...
>
> ... For describing our mental activity, we require, on one hand, an objectively given content to be placed in opposition to a perceiving subject, while, on the other hand, as is already implied in such an assertion, no sharp separation between object and subject can be maintained, since the perceiving subject also belongs to our mental content. From these circumstances follows not only the relative meaning of every concept, or rather of every word, the meaning depending upon our arbitrary choice of viewpoint, but also that we must, in general, be prepared to accept the fact that a complete elucidation of one and the same object may require diverse points of view which defy a unique description. ... The necessity of taking recourse to a complementary, or reciprocal, mode of description is perhaps most familiar to us from psychological problems.[9]

Bohr then draws the analogy between the "unity of our consciousness" and the "physical consequences of the quantum of action" which makes for the individuality of atomic processes. Using the term "emotion", to refer to the immediate subjective feeling of freedom, and "volition", to refer to that which is objectively described in an act of willing, he points to the "suggestive analogy" between these concepts as employed in the two different modes of describing the act of willing in psychology and the two modes necessary for describing an object in physics. Just as the incautious use of concepts as classically understood causes one to misunderstand the problem of wave–particle dualism as a dilemma about the nature of physical reality, so here "the problem of free will" is created by assuming that "one and the same" object is being described, first through reference to the subject's feeling of freedom ("emotions") and then through tracing a causal chain of objective acts of will ("volition"). Bohr argues that this use of complementary terms for describing consciousness is a feature of ordinary language where presumably different contexts imply different ways of drawing the distinction between observing subject and described phenomenal object:

> Actually, ordinary language, by its use of such words as thoughts and sentiments, admits [a] typical complementary relationship between conscious experiences implying a different placing of the the section line between the observing subject and the object on which attention is focused. ... In fact, the varying separation line between subject and object, characteristic of different conscious experiences is the clue to the consistent logical use of such contrasting notions as will, conscience and aspirations, each referring to equally important aspects of the human personality.[10]

Unfortunately, all Bohr gives us here is a "clue", but that clue should make it clear that he regards it as necessary to combine the mode of description of introspective, rational psychology with the mode of naturalistic, empirical psychology in order to present a complete, unambiguous description of all "equally important aspects of the human personality".

In 1929 when the Planck *Festschrift* article was written, Bohr was still liable to the confusion of speaking as though the observing process "disturbs" the object, the very confusion which, as we have seen, he cautions against in 1937. Thus, not surprisingly, at this time he carries over the same confusion in the psychological analogy: "The unavoidable influence on atomic phenomena caused by observing them here corresponds to the well-known change in the tinge of the psychological experiences which accompanies the direction of attention to one of their various elements."[11]

This reference to a "change of tinge" and the unity of consciousness suggests that Bohr's thoughts about psychology here were influenced by William James. Indeed, in the paragraph immediately following the previous passage, Bohr continues, "When considering the contrast between the feelings of free will, which governs psychic life, and the apparently uninterrupted causal chain of the accompanying physiological processes, the thought has, indeed *not eluded philosophers* that we may be concerned here with an unvisualizable relation of complementarity."[12] Bohr's wording here, so reminiscent of James' vocabulary in "The Stream of Consciousness", suggests that James may well be the "philosopher" who has caught the idea of complementarity in this context. Spinoza might also come to mind, but in his last interview, Bohr was asked pointedly what he thought of the epistemological analyses of Spinoza, Hume, and Kant, to which he replied that at the time he "felt these various questions were treated in an irrelevant manner" by these philosophers. As we saw in Chapter Two, the one philosopher who apparently excited his interest was James. Though the date of Bohr's reading of James is a matter of dispute, the wording of this essay suggests a familiarity with James' *Principles of Psychology* at least as early as 1929.

Bohr was naturally anxious that his comments in favor of a role for the concept of free will in describing consciousness were not understood as asserting the dualistic view of a causal influence between a nonmaterial "mind" and the physiological organism. Complementarity is relevant only in the description of *physical* interactions, not such presumed psycho-physical interactions of a dualistic metaphysics which are impossible to describe physically. Idiosyncratically, he appropriated the term "mysticism" to refer to such dualistic metaphysical doctrines. His concern with avoiding such a misinterpretation appears in his next paper, also written in 1929:

> ... the linkage of the atomic phenomena and their observation, elucidated by the quantum theory compel[s] us to exercise a caution in the use of our means of expression similar to that necessary in psychological problems where we continually come upon the difficulty of demarcating the objective content. Hoping that I do not expose myself to the misunderstanding that it is my intention to introduce a *mysticism* which is incompatible with the spirit of natural science, I may perhaps in this connection remind you of the peculiar parallelism between the renewed discussion of the principle of causality and the discussion of a free will which has persisted from earliest times. Just as freedom of the will is an experiential category of our psychic life, causality may be considered as a mode of perception by which we reduce our sense perceptions to order. At the same time, however, we are concerned with idealizations whose natural limitations are open to investigation and which depend upon one another in the sense that the feeling of volition and the demand for causality are equally indispensable elements in the relation between subject and object which forms the core of the problem of knowledge.[13]

In this passage, we can also note that Bohr does not mean to endorse the position taken by some physicists that the advent of indeterminacy at the quantum level destroys the causal chain on which the arguments for a biophysical determinism would rest. Indeed, he emphasizes the indispensability of a category of causality for the ordering of our psychic life, which the insistence on an absolute indeterminism would turn to chaos.

Several points emerging in this application of complementarity to psychological descriptions are not only typical of Bohr's analysis of complementary relationships in atomic physics, but also hallmarks of any complementaristic analysis of empirical knowledge in general.

First, what has most often been understood as a problem concerning the nature of reality (Are electrons really waves or particles? Is the will really free or determined?) is understood in a different manner as a problem in the use of concepts for describing some aspect of experience.

Second, two modes of description are required. In one the object is described as interacting directly with the subject or the observing system. In psychology this mode of description is introspective, allowing the subjective feeling of freedom to characterize the object; in physics it allows the application of the conservation principles. However, if one does not include a description of the whole phenomenon involving the agencies of observation such a description is ambiguous, because what is described is the whole phenomenon of interaction in which subject and object cannot be unambiguously distinguished. In psychology this is the actor describing his own decision to act, in physics it is the interaction between systems in the quantum domain and agencies of observation. Thus to make the description objective, *i.e.* to be able to describe the object unambiguously, a second mode of description must be combined with the first. In this mode the object is described as isolated from an observing interaction. But in so describing it, the descriptive terms must be understood as referring to an abstraction necessary for a causal description of the observing interaction, not a picture of an independently real object.

Third, any unambiguous description must necessarily distinguish between the observing agency and the observed object. This distinction may be drawn at any point, making possible describing the phenomenon as an observation of different phenomenal objects, each requiring a different description on a different "plane of objectivity". To avoid ambiguity, the description of each phenomenal object must specify how the distinction between observed object and agencies of observation has been drawn. Failure to do this renders the description liable to ambiguity due to the implicit but illicit shift of the subject/object distinction, thereby equivocating on the "object" thus described.

Fourth, failure to be sensitive to such ambiguities has given rise to the appearance of genuine metaphysical problems about the nature of reality which will disappear once the complementaristic analysis is employed. Problems such as wave–particle dualism or free will versus determinism are not metaphysical conflicts about the nature of reality. Instead these problems are confusions created by failing to realize that such different descriptions refer not to the same object but to complementary phenomena which only together provide an unambiguous description of the nature of the objects which give rise to these phenomena.

Fifth, the improvement of our understanding of nature made possible by adopting the framework of complementarity (and thereby revealing the deceptive character of such pseudo-metaphysical problems) occurs not through the invention of newer, more sophisticated concepts for describing experience, but through understanding the conditions required for the unambiguous employment of the descriptive concepts. This epistemological lesson, summed up in the reminder that we are actors as well as spectators in the drama of existence, results in that widening of our conceptual framework which Bohr calls a "rational generalization". In the following section we shall see how these same points also appear in Bohr's analysis of the description of biological phenomena.

3. Complementarity in Biology

In Chapter Two we noted that as a physiologist, Bohr's father was deeply interested in the dispute between mechanism and vitalism in biological descriptions, a burning issue in the late nineteenth century. There can be no doubt that Bohr considered his attempt to resolve this dispute to be in some fashion a payment of the intellectual debt he owed his father. We have repeated testimony that he derived great satisfaction over this use of complementarity. Indeed, of all of his discussions of complementarity outside of atomic physics, this application to biological problems was the most recurrent and the only one he worked out in any detail.

Bohr was very anxious to avoid appearing to support that version of vitalism which argued that organic systems possess a "vital stuff", some sort of nonphysical entity which inorganic objects do not possess and which can be described adequately only in terms of purposes. He warns sternly against any such *metaphysical* defense of the need for finalistic descriptions in biology, maintaining what he considers to be the naturalistic view that organic objects are no more than complex physical systems:

> ... this view, often known as vitalism, scarcely finds its proper expression in the old supposition that a peculiar vital force, quite unknown to physics, governs all organic life. I think we all agree with Newton that the real basis of science is the conviction that Nature under the

> same conditions will always exhibit the same regularities. Therefore, if we were able to push the analysis of the mechanism of living organisms as far as that of atomic phenomena, we should scarcely expect to find any features differing from the properties of inorganic matter.[14]

Just as his resolution of free-will versus determinism in psychology refused to countenance any mentalistic platform, so in biology Bohr's endorsement of a role for teleological modes of description does *not* involve abandoning a purely naturalistic approach to describing biological systems.

It is worth noting that just as in the case of psychology, Bohr believed that in biology also the lesson of complementarity had already been learned "more or less intuitively".[15] Starting from the observation that "any scientific explanation necessarily must consist in reducing the description of more complex phenomena to that of simpler ones", Bohr proceeded to point out that the progress of atomic physics has revealed "the unsuspected discovery of an essential limitation of the mechanical description of natural phenomena".[16] The limitation to which he refers here is not that of the uncertainty principle itself but rather that imposed by the need to accept the quantum postulate in order to describe the stability of the atomic system. As we saw in Chapter Three, Bohr's concern to account for atomic stability led him to adopt the quantum postulate in 1912.

Attributing the property of stability to an atomic system implies that, at that time, the system is regarded as isolated, not interacting with another system. No doubt in his later essays this concern with atomic stability fades from the picture, but it is apparently still on Bohr's mind, quite as one would expect, in his earlier essays on complementarity. In these essays, he seems to regard the atomic property of stability, which is destroyed by observing the system, as analogous to the organism's property of "life", which is destroyed by the observations necessary for a mechanistic description of its internal processes. On the one hand a mechanical description of the functioning of an organ within an organism is possible, but this requires that the organism be described as an isolated physical system. However, such an isolation precludes describing it as "living", for that concept refers to a property which can be manifested only through the organism's interactions with its surroundings, other physical systems. On the other hand, a description of the living organism is possible by describing the interaction of the organism with its environment, including the observing system. But since the observing system necessary for the observations which provide the empirical basis of the mechanistic description must now fall on the object side of the observer/observed distinction, the *object* described in the vitalistic mode of description which represents the organism as alive, is a *different object* from the one described in the mechanistic mode which reduces the description to the behavior of matter in space and time. Since the physical conditions necessary for the two modes are exclusive, they cannot be simultaneously employed and so must be used complementarily, but they must *both* be used for a complete description of a *living* organism.

This parallel between describing the stability of atomic structures and the life of an organism would be quite straightforward but for one crucial difficulty. We can understand this difficulty better by following how it emerged, and was eventually overcome, in a series of Bohr's essays on complementarity in biological descriptions. In his first essay on the topic, "Light and Life" written in 1933, he begins by noting that the complementaristic analysis is necessitated by the fact that in atomic processes the observing and observed systems form an interacting whole incapable of nonarbitrary subdivision.[17] He tries to forestall the misunderstanding produced by the claim that atomic processes are the physical foundation for biochemical, and hence vital, processes; this is not the point. Rather, he argues that a holistic analysis cannot in principle be reducible to an analysis of the components of a system which possesses a certain "individuality". Bohr describes the organic system as possessing an "organization" which is destroyed by any attempt at describing the physical processes of the component organs, due to the necessity for interacting with them in order to observe them in the way required for a mechanistic description. Thus he writes:

> An understanding of the essential characteristic of living beings must be sought, no doubt, in their peculiar organization, in which features that may be analyzed by the usual mechanics are interwoven with typically atomistic traits in a manner having no counterpart in inorganic matter. ... The question at issue, therefore, is whether some fundamental traits are still missing in the analysis of natural phenomena, before we can reach an understanding of life on the basis of physical experience.[18]

The "atomistic traits" to which he refers here are clearly not intended in the sense of the properties of atomic systems, but rather in the sense of "holistic traits" which cannot be defined at lower reductive levels of analysis. Nevertheless, there is an implicit confusion in this line of argument. As we have learned, the impossibility of describing separately the states of two interacting systems is a physical consequence of the quantum postulate in describing atomic processes. In the biological case it would seem that the impossibility of giving a mechanistic description of the processes inside an organism while that organism is alive is more like an *experimental* or *technical* impossibility which it is logically possible might be overcome by more subtle methods.

Of course quantum effects do play a role in biochemical processes at the molecular level, but this is not the basis of Bohr's application of complementarity to biological descriptions. The feature of "individuality" in biological systems which parallels that in atomic physics is the existence of *life* at the organic level. Bohr claims that the mechanistic framework is not limited in describing processes which go on within organisms, but he argues that the physical demand for an interaction with observing instruments at that level of observation, *i.e.*, the atomic level, physically is inconsistent with the maintenance of the organic functions of the organism, and hence with the organism manifesting the phenomenal display of life. For this reason, a description of a biological system cannot reduce it to a mechanistic system and still preserve reference to life. Thus Bohr writes:

> ... if we were able to push the analysis of the mechanism of living organisms as far as that of atomic phenomena, we should scarcely find any features differing from the properties of inorganic matter. ... however ... the conditions holding for biological and physical researches are not directly comparable, since the necessity of keeping the object of investigation alive imposes a restriction on the former which finds no counterpart in the latter. ... On this view, *the existence of life must be considered as an elementary fact that cannot be explained*, but must be taken as a starting point in biology, in a similar way as the quantum of action, which appears as an irrational element from the point of view of classical mechanical physics, taken together with the existence of elementary particles forms the foundation of atomic physics. The asserted impossibility of a physical or chemical explanation of the function peculiar to life would in this sense be analogous to the insufficiency of the mechanical analysis for the understanding of the stability of atoms.[19]

Unfortunately Bohr fails to note an apparent *disanalogy* in this account of biological description. Complementarity in atomic physics arises due to the empirical fact that in describing an observational interaction, the wholeness of the phenomenon of interaction implies that the distinction we draw between object and observing agencies is an arbitrary one. In biology the existence of life may be taken as an irreducible empirical fact analogous to the quantum of action, but it was only later that Bohr came to understand that there is no parallel demand that the interaction with the interior mechanisms of the organism in order to observe them forms an *indivisible whole* in a manner which makes any division between them arbitrary. Instead, the recourse to complementary descriptions occurs because we seek descriptions of biological phenomena on two distinct levels. First there is the level of the organism displaying vital behavior patterns in which its member organs function as part of this system, or, in Bohr's words, "organization", described in terms of functions and purposes. And there is the level which describes the mechanism that allows these organs to display such functions via a mechanistic, and ultimately biochemical, description reaching the quantum level. There is, no doubt, a great *practical* difficulty in keeping the organism alive while conducting researches on the biochemical functions taking place in its member cells, but no matter how *difficult* technically, this does not produce the sort of physical *impossibility* Bohr claims for the description of quantum mechanical systems.

However, Bohr supplements this first argument (that investigating the mechanical functions of the organs prohibits keeping the organism alive) with an indication of a second possible line of argument: "We cannot even tell which atoms belong to a living organism, since any vital function is accompanied by an exchange of material, whereby atoms are constantly taken up into and expelled from the organization which constitutes the living being."[20] Here Bohr's point more closely parallels the previous arguments from atomic physics. A strictly *mechanistic* unambiguous description of the behavior of organs requires that the physical system be *isolated* from interaction with other physical systems; however, the isolation of any biological organ from interaction with other organs of the physical system in its environment necessary to keep the organism alive would cause the death of the organism. Thus just as speaking of observing an isolated system in physics is literally a contradiction in terms which must be understood as an idealization necessary for defining the state of the object of description, so speaking of an *isolated living* organ or organism would appear to be equally literally a contradiction in terms. In fact, as long as it is alive, the organism must interact with the surrounding medium that supplies the needs of its vital processes in a way which renders arbitrary any distinction between the organism and those aspects of its environment (*e.g.*, food, air) on which its life depends.

The isolation demanded for a mechanistic description is isolation of the organism from its environment, with which it must nevertheless interact in order to remain alive. Thus the necessity for interaction is a *biological necessity* imposed by the nature of life. However, in physics the isolation required is incompatible with observing the object *at all*, for the observing instruments are themselves described as systems with which the object must interact for there to be any observation. In biology the isolation necessary for a mechanistic description of the organs within the organism is an isolation which is incompatible with maintaining the *life* of the organism because of its dependence on interaction with its surroundings to maintain its life. This "wholeness" is just as essential to describing the phenomena of life as the wholeness of observing system and observed object is to describing quantum phenomena. Thus in the mechanistic mode of description any sharp separation between organic system and environment is necessarily an arbitrary distinction, though one which must be made to provide a strictly mechanistic description of the organism. Of course the organism can be cut off from any interaction with its environment, but only at the cost of killing its display of the phenomena of life functions. Thus mechanistic and finalistic modes of description would seemingly be properly understood, not in the sense of opposed metaphysical doctrines about the sort of an entity which is the object of biological descriptions, but rather as distinct, but complementary, descriptions of the phenomenon of life, produced by separate ways of drawing the distinction between what is considered the organism and what is considered part of the environment within which that organism is found. Focusing on the vital processes of self-preservation and propagation, Bohr concludes that ...

> ... the essence of the analogy considered is the typical relation of complementarity existing between the subdivision required by a physical analysis and such characteristic biological phenomena as the self-preservation and propagation of individuals. It is due to this situation, in fact, that the concept of purpose, which is foreign to mechanical analysis, finds a certain field of application in problems where regard must be taken of the nature of life.[21]

Bohr's point is that describing an organism by defining the mechanistic states of its component organs is incompatible with the necessary physical condition for observing the object which biology seeks to describe, for that object is a system manifesting the property of life and all its concomitant processes. Thus in biology, as in physics, the two descriptive goals of "definition" and "observation" require different modes of description which describe distinct but complementary phenomenal manifestations of the object of biological description.

Bohr came back to the idea of the complementarity between mechanistic and finalistic descriptions in biology time and time again. In an article from

1937 we find him again rejecting the argument for the need for a vitalistic description on the basis of an *ontological* distinction between organic beings and physical objects. Instead, he insists, the problem of describing living systems requires being sensitive to the conditions required for the observations which yield the experiences to be described: "The general lesson of atomic theory suggests that the only way to reconcile the laws of physics with the concepts suited for a description of the phenomenon of life is to examine the essential difference in the conditions of the observation of physical and biological phenomena."[22] Here the term "biological phenomena" means phenomena in which life processes are manifested, not simply biochemical reactions between organic molecular structures, and the phrase "the conditions of the observation of ... biological phenomena" refers to the fact that to remain alive the organism must interact with other physical systems in its environment. The term "physical phenomena" here refers to the description of organic processes on purely mechanistic lines. In this article Bohr brings together both lines of argument which were not welded adequately in his earlier article, "Light and Life":

> Every experimental arrangement with which we could study the behavior of the atoms constituting an organism ... will exclude the possibility of maintaining the organism alive. The incessant exchange of matter which is inseparably connected with life will even imply the impossibility or regarding an organism as a well-defined system of material particles like the systems considered in any account of the ordinary physical and chemical properties of matter. In fact we are led to conceive the proper biological regularities as representing laws of nature complementary to those appropriate to the account of the properties of inanimate bodies. ... In this sense the existence of life itself should be considered, both as regards its definition and observation, as a basic postulate in biology, not susceptible to further analysis. ... It will be seen that such a viewpoint is equally removed from the extreme doctrines of mechanism and vitalism. On the one hand it condemns as irrelevant any comparison of living organisms with machines. ... On the other hand, it rejects as irrational all such attempts at introducing some kind of special biological laws inconsistent with well-established physical and chemical regularities. ...[23]

However, Bohr still does not seem to recognize that the problem of keeping the organism alive while trying to observe it in a way necessary for a mechanical description of its component organs is, as he has analyzed it, apparently disanalogous to the quantum situation because the difficulty is merely a technical, practical difficulty and does not depend on the physical conditions required for unambiguously applying descriptive terms.

In his 1955 article, "Physical Science and the Problem of Life", Bohr returns to this theme again, but now for the first time he emphasizes that the complementarity between mechanistic and finalistic modes of description is the consequence of the limitations on the use of descriptive concepts imposed by the physical situations required for the appearance of the phenomena to be described, *i.e.*, the phenomena of life. He begins by calling attention to a feature of Laplace's "well known conception of a world-machine", which was overlooked by Laplace and subsequent thinkers who have endorsed this view: "In this whole conception, which has, as is well known, played an important role in philosophical discussion, due attention is not paid to the presuppositions for the applicability of the concepts indispensable for the communication of experience."[24] Here he first seems to recognize explicitly the difference between two claims. The first is that as an *empirical* fact, it is "infeasible" to keep the organism alive while trying to describe the mechanical states of its component atoms. The second is that defining the state of the organism *as a purely mechanical system* requires physical conditions which are incompatible with the conditions necessary for the interaction of the organism with its surroundings essential to displaying the phenomena to which we refer when describing the organism as having the property of "life". Thus he writes:

> ... it must be stressed that an account, exhaustive in the sense of quantum physics, of all the continually exchanged atoms in the organism *not only is infeasible but would obviously require observational conditions incompatible with the display of life*. However, the lesson with respect to the role which the tools of observation play in defining the elementary physical concepts gives a clue to the logical applications of notions like purposiveness foreign to physics, but lending themselves so readily to the description of organic phenomena. Indeed, on this background it is evident that the attitudes termed mechanistic and finalistic do not present contradictory views on biological problems, but rather stress the mutually exhaustive observational conditions equally indispensable in our search for an ever richer description of life.[25]

Bohr does not mean to hold that a mechanistic description of the interaction between the organism and the surrounding systems is impossible. Quite to the contrary, just as in physics, such an account is not only possible, it is *necessary* for a complete description. His point is that in order to describe unambiguously such interactions mechanistically, the states of the interacting systems must be defined in isolation. The attempt to do this arbitrarily cuts off the organism from the systems necessary to support its life, thus destroying the physical conditions necessary for the display of the vital phenomena. Since the observing system and the observed organism do not form an indivisible whole (as they do in quantum interactions), it is possible to take into account the effects of the observational interaction:

> The basis for the complementary mode of description in biology is not connected with the problems of controlling the interaction between object and measuring tool, already taken into account in chemical kinetics, but with the practically inexhaustible complexity of the organism.[26]

What cannot be done, due to the "practically inexhaustible complexity of the organism", is to define its state apart from its living environment. But Bohr's expression is misleading in suggesting that the impossibility of discarding vitalism in favor of mechanism is a consequence of the *practical* difficulties involved in an attempted mechanistic analysis of a system as complex as even the simplest organism. Such a task may well be *practically* impossible, but the necessity for making the finalistic mode of description complementary to the mechanistic arises in the physical incompatibility between the conditions necessary for the unambiguous application of mechanistic descriptive concepts (isolation) and the conditions necessary for the display of life phenomena (interaction with surroundings).

In these last lectures Bohr seems to realize that complementary descriptions in biology are necessary because biology is by definition the description of systems exhibiting the phenomena of life. Thus it seems he sees progress in biology as advancing on two distinct complementary fronts. One front is the level of biochemical descriptions of the physical and chemical mechanisms responsible for the various phenomena manifested at the subcellular level; the other is on the level of the organism as a whole and its interacting role within larger systems essential to maintaining its life.

By this point Bohr had apparently recognized the difficulty with his earlier claim that the interactions necessary for those observations which would permit a mechanistic description of vital processes would *as a matter of empirical fact* cause the death of the organism. Instead, now he argues that complementary descriptions are necessary in biology because the physical conditions necessary for the appearance of the phenomenon of organic life require physical interactions with those systems necessary to keep the organism alive. Thus if we use the term "life" in a description, we cannot, as a consequence of the presuppositions necessary for the application of that term, physically isolate the system as is required for a mechanistic mode of description at the atomic level:

> This situation can hardly be regarded as being of temporary character, but rather would seem to be inherently connected with the way in which our whole conceptual framework has developed from serving the more primitive needs of daily life to coping with growth of knowledge gained by systematic scientific research. Thus as long as the word "life" is retained for practical or epistemological reasons, the dual approach in biology will surely persist.[27]

Though the demand for complementary descriptions is contingent upon the continued use of the descriptive concept of "life", Bohr makes it clear that this situation is not an unfortunate half-way house to be left behind should improved experimental techniques make possible a purely mechanistic description, any more than the wave–particle dualism of atomic physics marks an incomplete description to be left behind when one description proves victorious over the other.

In his very last lecture, delivered at Cologne, in honor of his one-time student and pioneer molecular biologist, Max Delbrück, Bohr returned for a final time to the theme of complementarity in biology in the essay, "Light and Life Revisited", partially edited for posthumous publication. Here he comments on his former speculation "that the very existence of life might be taken as a basic fact in biology in the same sense as the quantum of action has to be regarded in atomic physics as a fundamental element irreducible to classical physical concepts".[28] He indicates implicitly that he has abandoned his earlier position that keeping the organism alive is inconsistent with "a full atomic account" and concentrates on the fact that "the task of biology cannot be that of accounting for the fate of each of the innumerable atoms permanently or temporarily included in a living organism".[29] He argues instead that "structure" as described mechanistically and "function" as described teleologically represent complementary manifestations of organic phenomena. The descriptions are complementary because a teleological description of the functioning of the organs in an organism provides the starting point for a mechanical analysis of that physical structure which makes possible such functioning. However, in the mechanistic description of the organism the interaction between it and its surroundings must be left out of account in order to define the state of the system as physically isolated. The conceptual distinction between describing structures mechanistically and their functioning teleologically reflects the different modes of description which are interdependent in complementarity:

> In the study of regulatory biological mechanisms the situation is rather that no sharp distinction can be made between the detailed construction of these mechanisms and the functions they fulfill in upholding the life of the whole organism.[30]

Recalling his traditional stress on the complementary relationship between the two modes of description in physics, Bohr concludes that a mechanistic description of such regulatory processes in physiology necessitates a prior teleological description in order that a mechanical analysis can be attempted:

> Indeed, many terms used in practical physiology reflect a procedure of research in which, starting from the recognition of the functional role of parts of the organism, one aims at a physical and chemical account of their finer structures and of the processes in which they are involved. Surely as long as for practical or epistemological reasons one speaks of life, such teleological terms will be used in complementing the terminology of molecular biology.[31]

In these last extensions of complementarity into biology we are again struck by Bohr's unwavering allegiance to that central epistemological lesson of complementarity: we advance our understanding of nature, rather than merely quantitatively increase it, only to the extent that we gain fuller recognition of how our concepts function in describing experience. Progress in science does not require narrowing down our descriptive vocabulary to a few correct terms. Instead, real advance is made possible only by constantly widening that conceptual framework with which we first approach our experience in everyday life. In such a fashion we must revise our understanding of the presuppositions governing the use of those concepts, thereby clearing room for the description of new phenomena and stimulating that sense of wonder in which all natural philosophy begins.

CHAPTER SEVEN

COMPLEMENTARITY AND THE NATURE OF EMPIRICAL KNOWLEDGE

Although Bohr's failure to convert Einstein was no doubt a significant disappointment to him personally, the nature of their disagreement was such that it could be put to profitable issue, for both men spoke the language of physics. However, Bohr's failure to attract any notable audience among philosophers was for him a cause of bewilderment and frustration. The simple fact was that Bohr and the philosophers hardly spoke the same language. Thus we recall Bohr's telling comment in his last interview: "I think that it would be reasonable to say that no man who is called a philosopher really understands what is meant by the complementary descriptions."[1]

As we turn now to the specifically philosophical questions with which complementarity is concerned, it will be helpful to consider in general the central issue in the philosophy of science over which communication broke down between Bohr and "the philosophers". I refer here to the issue which divides those who hold that scientific theories are attempts to describe the phenomena we experience as the empirical consequence of the behavior of objects which lie behind these phenomena from those who do not. The former hold the view called "realism"; their opponents are "anti-realists". However, it is hardly the case that all defenders of realism in science need defend the same form of realism. Indeed, I have already indicated in part that the issue which separated Einstein and Bohr was between two alternative forms of realism, although it has been common to characterize their difference as one between realism and anti-realism. Thus it will be convenient to distinguish "classical realism" as the view which results from a realistic reading of classical physics and contrast this position to the form of realism inherent in complementarity. The classical realist holds that the success of theories operating in the framework of classical physics legitimizes the claim that phenomena are the consequence of the behavior of objects having properties corresponding to the terms used to characterize physical systems in classical mechanics. As we have seen, on this classical realist view, the concept of the state of the system defined in terms of the classical mechanical parameters may be regarded as picturing the properties possessed by an independently real object which causes the phenomena that confirm the theoretical representation. In this chapter and the next I will argue that Bohr defends a form of realism which denies this claim but offers the possibility for an alternative form of knowledge of a domain lying behind the phenomena.

As there are various forms of realism, so there are different forms of anti-realism. The dominant one during Bohr's career was that of "instrumentalism", the view that theoretical terms serve only as constructs enabling correct inferences to predictions concerning phenomena observed in specifiable circumstances. Many defenders of anti-realism also hold the view of "phenomenalism", the assertion that the only reality of which we can form an idea with any content is that of phenomena, and that therefore statements about a reality lying behind the phenomena are meaningless. Both of these views have been imputed to Bohr quite incorrectly, as I will try to show.

Although the issue of realism versus anti-realism provides a revealing philosophical background for understanding complementarity, Bohr did not see the issue between himself and the philosophers in terms of the ontological question of the nature of physical reality. For him the battle was joined over whether or not complementarity provided an "objective" description of phenomena. Thus in his last interview he complained that the philosophers "did not see that it was an objective description, and that it was the only possible description. So therefore the relationship between scientists and philosophers was of a very curious kind."[2] As this comment indicates, Bohr understood the communications breakdown between himself and the philosophers in terms of the epistemological issue of the requirements for an "objective description". Thus that will be the central focus of this chapter on the epistemological significance of complementarity. But before we turn directly to that matter, we need consider first how complementarity is related to the theory of empirical knowledge and then why it has seemed that Bohr's account of empirical knowledge in the quantum domain is "subjectivistic".

1. The Relationship between Complementarity and Epistemology

As a conceptual framework for describing nature, complementarity claims to be a rational generalization of the framework of classical mechanism. Mechanism is not an epistemology *per se*, and so, by analogy, complementarity is not either. Nevertheless, it is an historical fact that the advent and subsequent dominance of mechanism had immediate repercussions on epistemology and that the success of mechanistic science was considered a major datum for the traditions of rationalism and empiricism, as well as Kant's critical philosophy. Thus it is natural to ask, if complementarity is a general conceptual framework within which we are to understand the description of nature in natural science, what does it tell us about the nature of scientific knowledge?

As the framework of classical physics, mechanism provided a directive to the natural scientist to look for theories and descriptions of a certain kind, those which would be assessed as "adequate" or "acceptable" descriptions within the mechanistic framework. Mechanism thus presents an "ideal" for the description of nature by stipulating what sorts of accounts of phenomena will be regarded as acceptable claims to scientific knowledge. Such an ideal gives a standard against which prospective theoretical descriptions may be measured. In this way mechanism not only provides a framework of concepts interrelated in certain ways such that they have specific meanings in the description of nature, but also it serves a directive function in orienting the progress of science towards the projected goal of describing an observed phenomenon in accord with its ideal of description.

As we have seen, the development of quantum mechanics was originally conditioned by the ideal of mechanism, but measured against the standards of the mechanistic framework, standards which Einstein for one was not wont to regard as changeable, quantum theory must be rejected as incomplete. These standards imply that a complete, objective description is one which determines the properties possessed by an independently real physical object by representing that object as a physical system isolated from any observational interaction existing in a well-defined classical mechanical state. If these properties are considered to be represented in the theory by the classical state parameters, then it follows that the inability of quantum mechanics to represent the object of its descriptions as in such classically well-defined states, means that it does not fulfill what the classical framework expects of an acceptable theory.

Bohr designed complementarity to change the standards for an "ideal" description so that quantum theory would be evaluated as presenting an acceptable, consistent, and complete description of atomic phenomena. Therefore the switch in conceptual frameworks from mechanism to complementarity implies a change in the standards of what is defined as acceptable science. Thus if natural science is taken as a paradigm of human knowledge, as it has been in most Western epistemology since the Enlightenment, changing the descriptive ideals of natural science clearly implies changing a fundamental datum with which any epistemology of empirical knowledge must come to terms. This recognition should be the consequence of Bohr's epistemological lesson.

The classical framework and complementarity do not differ verbally in their approbation of the products of science as "objective descriptions". However complementarity changes the significance of the term such that what Bohr means when he claims objectivity for the complementary descriptions is significantly different from what Einstein, for example, would demand.

Bohr begins his epistemological reflections with the recognition that empirical knowledge must ultimately be based upon experiences which are the property of a perceiving subject, but in scientific knowledge we proceed from this subjective starting point to an objective account of the experienced phenomena. Thus for Bohr "objectivity" is a property of the descriptions of phenomena which science provides. The common perception of classical mechanism as a program for eliminating purposive and finalistic descriptions of natural phenomena is a consequence of the fact that such descriptions were held to result from importing *subjective* categories into the description of physical phenomena. Nevertheless, the empirical reference of the mechanistic concepts of spatial and temporal locus to aspects of perceived phenomena have equally much a *subjective* basis as the feelings of desire and purpose which premodern science employed to characterize the objects of its descriptions. However, they provide the terms for an "objective description" because they refer to what is directly observable to *any* normal experiencing subject *and* can be communicated by unambiguous measurements expressed in the language of mathematics.

Although the grounds for the acceptance of a theory is ultimately dependent on the empirical reference of the spatio-temporal concepts to specific aspects of human perceptual experience, classically it was possible to hold that the unambiguous descriptions which these terms made possible referred to a reality behind the experienced phenomena. Thus to a classical realist, the "objectivity" of mechanical descriptions implied much more than that they were unambiguously communicable. Objectivity could be grounded in a reality behind the phenomena because the terms defining the state of an unobserved isolated system could be consistently held to refer to the properties of an independent reality. If one could assume some sort of causal link between the mental and the physical, it was possible to consider the properties of observed phenomena to which the terms referred as the mechanical causal effects of properties possessed by an independent reality interacting with the perceiving subject's physical sense organs. Thus in spite of the subjective status of such perceptions, the fact that *any* normal subject always had these perceptions, under suitably controlled circumstances, assured the consistency of maintaining that the property perceived as belonging to the phenomenal object was grounded in the property of an objectively real object the nature of which it is the task of science to describe.

The epistemological difficulties of this position are a matter of record in the history of philosophy. With Berkeley's and Hume's empiricist critiques of the idea of an independent physical reality, experience was made to be free-floating from any dependence on the properties of supposed independently real objects lying behind the phenomena. Thus the *ontological* grounds for the classical claim to the objectivity of its descriptions was eliminated. The alternatives were to embrace either a *phenomenalism* in which "objectivity" becomes in fact, if not in word, replaced by "inter-subjectivity" or an *idealism* in which the objective status given to various terms of descriptive significance is provided by the allegedly necessary cognitive operations of the subject. Thus if complementarity is intended to be merely a principle within mechanism, and for this reason to be evaluated by reference to the ideals of that framework, it is not surprising that it has been interpreted as supporting either phenomenalism or an idealist defense of the objectivity of specific descriptive concepts. Indeed, many writers on complementarity have tended to regard Bohr as either a phenomenalist or as a quasi-Kantian subjectivist, or as indecisively suspended between the two.[3] But we have seen that complementarity is intended to replace the mechanistic framework, hence it is not to be judged by the ideals of description laid down by mechanism. For this reason, such readings of complementarity rest on a misunderstanding.

2. *Subjectivity and the Description of Experience*

How can a philosophical account of scientific knowledge reconcile the empiricist conviction that ultimately individual subjective experiences are the foundation of all knowledge of nature with the professed goal of objectively describing nature? This is a crucial issue with which the epistemology of science must reckon, and it is one with which Bohr and complementarity are vitally concerned. Classical physics had an answer, but its answer was based on a presupposition that the quantum revolution requires discarding. Thus a reanalysis of the concept of objectivity and a radical redefinition of this criterion of scientific knowledge form the philosophical heart of complementarity.

Bohr announces his empiricism with the very first sentence of *Atomic Theory and the Description of Nature*, written in 1929: "The task of science is both to extend the range of our experience and reduce it to order."[4] That he never wavered from this goal of what he liked to call "analysis and synthesis" is shown by the fact that almost exactly the same sentence occurs in "Atoms and Human Knowledge" written twenty-six years later: "The goal of science is to augment and order our experience."[5]

Bohr is very much aware that the requirement of objectivity imposed on the scientific description of nature is in paradoxical contrast to the subjective status of the experiences on which such a scientific description is based: "Yet occasionally just this 'objectivity' of physical observations becomes particularly suited to emphasize the subjective character of all experience."[6] Here Bohr clearly regards a "physical observation" as an objectively given datum. As we have seen, it is described as a physical interaction between object systems and observing instruments. However, in order to count as an observation, the experienced "phenomenon" (in Bohr's later sense) of that interaction must be described unambiguously. The basis for this description is the experimenter's experience of the phenomenon. That experienced phenomenon, like any phenomenon, is an experience of "everyday objects" in space and time. Thus it must be described by the usual terms, suitably refined, for describing everyday objects. In this way the "objectivity of physical observations" emphasizes the "subjective character of all experience".

Unfortunately, this reference to "subjectivity" was reckless, as Bohr came to realize from the persistent attempts of hostile critics to interpret complementarity subjectivistically. Thus in later years virtually every essay contains a disclaimer of any subjective intentions, for example, "the decisive point is that in neither case [*i.e.*, in neither quantum physics nor in relativity] does the appropriate widening of our conceptual framework imply any appeal to an observing subject, which would hinder unambiguous communication of experience".[7] While experience is of course always experience *of a subject*, the phenomenon which is experienced is an objective event. The field of all such phenomena constitutes "nature". This is the datum to which Bohr refers in speaking of science as "the description of nature".

Nevertheless the classical tendency to regard experience as subjective misleads one into reading complementarity as endorsing a subjective foundation for science. Bohr is explicitly aware of this temptation:

> In view of the influence of the mechanical conception of nature on philosophical thinking, it is understandable that one has sometimes seen in the notion of complementarity a reference to the subjective observer, incompatible with the objectivity of scientific description. Of course in every field of experience we must retain a sharp distinction between the observer and the *content of the observations*, but we must realize that the discovery of the quantum of action has thrown new light on the very foundation of the description of nature and revealed hitherto unnoticed presuppositions to the rational use of the concepts on which the communication of experience rests. In quantum physics, as we have seen, an account of the functioning of the measuring instruments is indispensable to the definition of the phenomena and we must, so to speak, distinguish between subject and object in such a way that each single case secures the unambiguous application of the elementary physical concepts used in the description.[8]

In the previous chapter we noted that Bohr believed that this shifting of the distinction between subject and object also appears to cause problems in psychology and biology. Thus in the essay from which the above passage is taken, he continues by pointing out that, "While in the mechanical conception of nature the subject–object distinction was fixed, room is provided for a wider description through the recognition that the consequent use of our concepts requires different placings of such a separation."[9]

These comments indicate that the task of securing the objectivity of empirical knowledge is presumed to refer to the *description* of experience, or in other words, what happens *after* the subject has the experience rather than how experience originates. Bohr avoids grounding objectivity in either the object's contribution or the subject's contribution to the *formation* of experience. His view is that objectivity is secured by understanding the proper use of descriptive concepts, and thus that it is a matter of how we *describe* an experienced phenomenon, rather than how experience originates, which makes for objective knowledge. This view appears in his comment that "in complementary descriptions all subjectivity is avoided by proper attention to the circumstances required for the well-defined use of elementary concepts".[10]

Bohr does not intend to treat the problem of objectivity *vis à vis* the origin of experience in the relation between ontologically distinct orders of an objective natural world and a subjective conscious mind. But neither does his insistence that the problem of objectivity is concerned with the *description* of experience entail any reductionist attempt to describe consciousness on a purely physicalistic basis. In the previous chapter we saw that Bohr holds that psychological phenomena require complementary descriptions and that the mentalistic language of conscious self-awareness is complementary to the physicalistic description of the same psychological object. Thus whether consciousness is regarded as reducible to a physical process (described in terms of a mechanistic analysis of the interactions of material substances) or as a mental presentation of ideas in a subjective consciousness is irrelevant to complementarity. What concerns Bohr is not the interplay of distinct ontological orders (as in the traditional philosophical characterization of the relation between subject and object in experience), but the necessity for describing experienced phenomena unambiguously.

A subjectivistic interpretation of quantum theory may well be possible, but as we know from Bohr's frequent attempts to ward off this interpretation, it is not that of complementarity. Such a reading of complementarity is clearly incorrect, for Bohr repeatedly emphasized that in no case is a *consciousness* one of the "systems" which enter into the creation of the experienced phenomena which physics describes as interactions between *physical* systems. His whole point in stressing that observation is interaction is that this observation must be describable as a *physical* interaction in terms of the classical concepts so that the description can be communicated unambiguously. If a human consciousness were one of the "systems" related in the "observational interaction", then a description of that interaction would involve a mental state that could not be described unambiguously in the concepts of classical physics, thus rendering such a description ambiguous and so "subjective". Furthermore, Bohr's whole case rests on the quantum postulate which expresses an empirical discovery about *physical* interactions. If he did not intend the observing interaction to be a purely physical one, then all of his arguments to show how complementarity is based on the quantum postulate would be completely irrelevant. Thus "observation" for Bohr does not involve any mysterious mental/physical interaction; in order to describe an observation in physics, both systems which interact to produce that observation must be capable of being described as purely physical systems.

Bohr makes emphatic his claim that there is no need for a new *observational* language in quantum physics. In other words, since classical descriptive concepts have unambiguous empirical reference for describing phenomena, quantum mechanical experiments present no need to be described in any observational language other than that of classical physics:

> Notwithstanding the power of quantum mechanics as a means of ordering an immense amount of evidence regarding atomic phenomena, its departure from accustomed demands of causal explanation has naturally given rise to the question whether we are here concerned with an exhaustive description of experience. The answer to this question evidently calls for a closer examination of the conditions for the unambiguous use of the concepts of classical physics in the analysis of atomic phenomena. The decisive point is to recognize that the description of the experimental arrangement and the recording of observations must be given in plain language, suitably refined by the usual physical terminology. This is a simple logical demand, since by the word "experiment" we can only mean a procedure regarding which we are able to communicate to others what we have done and what we have learnt.[11]

Bohr's point is that we do not need two descriptive languages, one for describing ordinary phenomena and the other for describing "quantum phenomena". Bohr uses the phrase "atomic phenomena" or "quantum phenomena" to refer to those experienced phenomena which are described as observations confirming quantum mechanics as an acceptable atomic theory. However, such phenomena do not differ generically from other phenomena. Thus all experienced phenomena can be described by the same language, the language suitable for the unambiguous communication of a description of experience.

Though experienced phenomena *can* be described in the classical physical language, the quantum theory's "departure from accustomed demands of causal explanation" causes one to wonder whether another language could be developed for describing phenomena which would not be thus restricted by the quantum postulate. Consequently it is essential to Bohr to maintain that the "plain language" of classical physics (suitably refined) is *the* language for describing phenomena. In fact it would be more revealing to say the reverse: the language developed for the unambiguous description of phenomena is the language of classical physics. Bohr argues that there is no need for a new language which would eliminate the complementary combination of two modes of description, because the classical language is already fully determinate; *i.e.*, it is capable of expressing everything that can be said about phenomena. There is no element in an observed phenomenon which cannot be unambiguously communicated in the classical language. Thus any attempt to develop another language which would make possible a more "complete" description is a futile search for a description which would convey more than a fully determinate description of the phenomenon. If we accept the quantum postulate as axiomatic, and the consequent need to use complementary modes of description, then we must accept that there is no lack of determination which the invention of a new set of descriptive terms would remove.

Speaking of "observing" a "microsystem" or an "atomic system" may be philosophically misleading if we confuse Bohr's use of "observing" with his use of "experiencing". The experimenter "experiences" the phenomena which confirm his theory. Some of these phenomena may be described as "observations" of atomic systems. Since the classical descriptive terms have unambiguous reference only for describing phenomenal objects, and since "atomic system" cannot refer to a phenomenal object, the classical descriptive terms cannot be used to describe atomic systems. Therefore, if any description of an atomic system is at all possible, such a description cannot be unambiguously communicated through the concepts used for describing phenomena. However, the only knowledge of such objects open to the empirical scientist is through describing those phenomena which provide the grounds for confirming the theoretical representation of the experienced phenomena as arising in the interaction between the observing instruments and an independently existing atomic system. But these phenomena which form the corroborating evidence for the quantum theoretical description are ontologically no different from those which serve to corroborate any empirical theory.

Once we understand the description of nature within the framework of complementarity, it is clear that its epistemological orientation is radically different from the "representationalism" underlying the realistic interpretation of the classical framework. According to this view the task of science as *objective* knowledge is to form a picture which represents how objects are apart from subjective human experience of them. Bohr dramatically alters this characterization of the description of nature. "Observer" and "object" become *categories of description* having precise meaning only in the *context of a particular description* of an experienced phenomenon. Apart from a particular descriptive context in which this distinction is specified, these terms are *ambiguous*, owing to the lack of specification of the requisite conditions for their application.[12]

The description of nature requires the description of experienced phenomena, but these phenomena are not regarded as a *representation* of something more fundamental, an independent physical reality. However, some phenomena can be treated as "observations" which are produced by interactions with objects not themselves present in the experienced phenomena.

Thus the descriptions of experimental phenomena as physical interactions will require reference to atomic systems as objects "behind" the phenomena, in the sense of that which produces the phenomena, but the phenomena experienced are not themselves "representations" of these objects. Although phenomena described as observations provide the foundation for accepting or rejecting particular proposed theories, "observations" cannot be "independent" of such descriptions, as was presupposed in the interpretation of science of traditional empiricism, because in order for a phenomenon to be relevant for the acceptance of a scientific description it has to be described as an observation of some determinate phenomenal object. Which phenomenal object is described depends upon the distinction between observing system and observed object drawn in a particular description provided in the context of a particular conceptual framework. At least with respect to those interactions which must be described as having a feature of "individuality", failure to heed this fact will produce an ambiguity in scientific description and an improper understanding of the use of descriptive concepts in science. However, the nature of the experienced phenomenon does give to observations the particular characteristics they do have, such that the empirical reference of descriptive terms is grounded in "nature", as the field of all experienced phenomena, independently of the particular theoretical description in which a concept is employed.

To be sure, Bohr's view of science as concerned with the description of "nature" does not refer to a domain of independently real physical objects. But it is incorrect to infer from this fact the conclusion that complementarity abandons any attempt to understand the nature of a world that exists "external" to human experience. The next chapter will focus on the ontological status of such an independent physical reality and the means by which we may understand it. My present point is that Bohr does intend to restrict the use of the "classical concepts" from describing a transphenomenal world. Thus his position includes two claims: a) in describing phenomena, including so-called "atomic phenomena" which differ in no way *qua* phenomena from any other phenomena, the empirical reference of classical concepts given in experience guarantees that their use is consistent and unambiguous as long as the description of the phenomenon refers to the whole experimental situation described as the observational interaction; and b) since those classical concepts are thus defined only for the *description* of *phenomena*, there are no grounds for holding that such concepts can meaningfully refer to or "represent" the properties of entities as they exist externally to our observations of them.

The classical position required that an independent reality possessed properties corresponding to the theoretical parameters through which the state of the system is defined and causing the properties of the experienced phenomena that confirm the theoretical representation. The experienced phenomenon, then, is the bridge between a theoretical representation and the properties of an independently real world. Such a bridge had the appearance of solidity as long as the epistemological realism and representationalistic theory of experience of the classical framework went uncriticized. But when these views were seen to be empirically unfounded, the bridge was left without any support.

Nevertheless, the belief that this bridge was well supported persisted because it entailed two very useful consequences. First, if the independently real world possesses properties "represented" by the classical mechanical state parameters, then the theoretical definition of the state of the isolated system could be regarded as presenting a picture of the properties of a real world existing apart from our observations of it. Secondly, it enabled classical physics to assume that the same descriptive concepts which made possible an unambiguous description of experienced phenomena could be used to describe an independent reality. In this way it was possible to "visualize" or "picture" what the world "looks like" when we are not looking at it. This theoretical reconstruction of physical objects as they would appear to an observer "detached" from any causal interaction with them then became the ideal for an *objective* description.

Bohr's epistemological lesson requires relearning how concepts apply to the description of nature. If the spatio-temporal parameters of theory are assumed to correspond to the properties of an independent reality, then the fact that we use both wave and particle pictures would seem to make the description ambiguous, for it does not tell us what sort of entity the object is. Consequently, Bohr holds that when the spatio-temporal concepts are used to predict the outcome of a given experimental interaction, they refer not to the properties of an independent reality, but only to properties of the experienced phenomenon. The physical conditions necessary for describing an observed object require that in the resulting description, the classical terms are predicates applying only to the object as experienced, the phenomenal object. Without that observational interaction, the classical concepts could have no empirical reference as is required for experimental confirmation of theory. By the same token, the physical condition of isolation necessary for a theoretical prediction necessitates that in the quantum formalism the state of an isolated system refers only to an abstraction. In this sense the representations of the formalism are indeed "inference tickets". Thus, were *this* concept taken to provide the total content of our knowledge of the properties of a physical system which is being described through the quantum mechanical formalism, as is assumed in the classical framework, then complementarity would indeed endorse instrumentalism. However, as we shall see in the next chapter, the complementarity of the two modes of description precludes taking this concept in just such a sense.

3. The Ideal of Objectivity

The fundamental question which epistemology must address is how to formulate a description of experience meeting the criterion of objectivity. Traditionally a scientific description was held to be "objective" because it was believed that it determined the properties possessed by the object as it exists apart from any observational determination of those properties. The methodology of science was built around the goal of developing a means for determining such properties and at the same time eliminating any element which arises from the role of the observing system in acquiring that empirical knowledge necessary for confirming a scientific theory. Bohr argued that this understanding of the objectivity of a scientific description of nature must be revised.

For Bohr the objectivity of scientific knowledge is grounded in the concepts used for the description of experience. These concepts secure the objectivity of scientific description not by playing any role in the origin or formation of experience, but rather by guaranteeing that communications expressed in terms of such concepts are unambiguous.

> Every scientist, however, is constantly confronted with the problem of objective description of experience, by which we mean unambiguous communication. Our basic tool is, of course, plain language. ... We shall not be concerned here with the origins of such language, but with its scope in scientific communication, and especially with the problem of how objectivity may be retained during the growth of experience beyond the events of daily life.
>
> The main point is to realize that all knowledge presents itself within a conceptual framework adapted to account for previous experience and that any such frame may prove too narrow to comprehend new experiences. ...[13]

As Bohr came to understand natural science, the objectivity of a description cannot be secured by separating the allegedly "objective" properties possessed by an independent reality from the "subjective" properties which exist only in relation to an experiencing subject. By the definition of "objectivity" in complementarity, the scientist's goal must be to develop a conceptual scheme or framework for describing phenomena in a way which can be communicated unambiguously. Once the quantum postulate is accepted, it becomes necessary to understand that any description of a phenomenon is the description of a physical interaction in which the distinction between observed object and observing system is made for the purpose of unambiguously describing the interaction as an observation of a specific phenomenal object. To secure the unambiguity of the communication of the result of this observation, it is necessary that the description includes a precise specification of the whole physical situation in which the interaction (the observation) occurs.

Since the task of science is to develop a framework for consistently describing all known natural phenomena which permits unambiguous communication, if a framework becomes inadequate as science explores new aspects of nature, descriptions which could once be regarded as unambiguous may no longer be so. This is precisely what has happened with wave–particle dualism in the classical framework. In describing the motion of everyday objects by reference to the trajectories of particles moving in space through time, a realistic interpretation of the Newtonian description could clearly consistently hold that the objects thus described were entities corresponding to the theoretical representation. But if this framework is used to describe atomic systems, say electrons, then exactly what we mean by describing the electron as a particle moving in a spatial trajectory cannot mean what it used to mean because we also describe other phenomena by representing such electrons as waves. Since we no longer know what is meant by describing nature through such a representation, the description becomes "ambiguous" in Bohr's sense of the word. Bohr perceives the paradoxical nature of such quantum mechanical descriptions as essentially the consequence of such "ambiguity". Complementarity is designed to remove that ambiguity, hence to secure an "objective description".

In other words, Bohr's epistemological lesson teaches that on the presuppositions of the classical framework, descriptions of phenomena expressed in terms of the classical mechanical concepts could be considered unambiguous. But the exploration of the atomic domain uncovered new phenomena dictating the need to adopt a new presupposition about these phenomena, that expressed by the quantum postulate. This change in the classical presuppositions then renders descriptions expressed in classical terms "ambiguous". Consequently, a new, more general framework (but not new concepts) is needed to restore the unambiguity of descriptions of those phenomena supporting quantum theory. Once this new framework revises the presuppositions governing the use of concepts, the objectivity of scientific descriptions of phenomena will be regained.

In order to follow why Bohr believes complementarity removes such ambiguity, we must remind ourselves that in basing the description of nature on *observation*, science becomes committed to accounting for how that observation reveals information concerning the observed object through interacting with it. An objective science must develop a way for describing this interaction such that its results can be communicated unambiguously. For this purpose the indispensable role of the classical mechanical concepts is that they establish a way for describing the observing instruments as revealing the results of the observational interaction in the precise language of mathematics. This point is of course common to both classical and complementaristic accounts. But the unambiguous character of mathematical language is of use only if the mathematically defined parameters can be given empirical significance to measurable aspects of experience. For Bohr the spatio-temporal concepts are irreplaceable for describing observation because the nature of experience is such that, as a matter of empirical fact, these concepts do have an unambiguous reference to the experience of the measuring instruments as everyday objects in space and time. Because of the empirical reference of the spatio-temporal concepts used in expressing observational results, a role which remains unchanged in the switch from classical to quantum mechanics, use of these "classical" concepts makes the description unambiguous. It was precisely because the objectivity of scientific description was for Bohr essential that he refused to abandon the classical concepts.

But in order to describe the observational interaction as a causal process, which must be done if that interaction is to be interpreted as an observation of the object system, it is also necessary to be able to apply the classical dynamical conservation principles. To do this one must have a theoretical representation of the system isolated from the interaction. However from the quantum formalism it is impossible to derive a classical mechanical state of the system that would enable us to picture the object apart from observation. Therefore we now recognize the theoretical representation of the isolated system as an abstraction, not a "picture" of a concrete real object.

The shift from the classical framework to complementarity requires changing the role of experienced phenomena as the empirical basis of the scientific description of nature. Classically, these phenomena served as the empirical link connecting an independent physical reality with the theoretical representation of that reality. The classical mechanical state parameters in terms of which the theoretical representation is defined have empirical significance by being made to refer to specific observable properties of phenomena. On the classical realist supposition that it is the task of science to describe the properties of an independently existing physical reality, the objectivity of descriptions using the classical concepts was based on the claim that these observable properties of phenomena are the causal effects of the properties of such an independent physical reality. Thus from the classical viewpoint when wave and particle concepts are used to refer to "radiation in free space or isolated material particles", the resulting "pictures" can be consistently regarded as representing the properties of independently real objects. As we have learned, in order to avoid the contradiction of wave–particle dualism, complementarity cannot accept this understanding of theory. Instead, Bohr argues that such representations are "abstractions" which serve an indispensable role in allowing a phenomenon to be described as an interaction between observing systems and atomic systems, but which cannot picture the properties of an independent reality.

The classical framework made the notion of an isolated system the basis of its claim for the objectivity of its description. However, its representation of an isolated system is regarded as objective because it is assumed to allow "visualization" of the properties of an *independent* physical reality in terms of the classical mechanical concepts which have empirical significance only by referring to observed properties of *phenomenal* objects. Complementarity accepts no such assumption. Instead it acknowledges the theoretical representation of the isolated system as an abstraction but calls attention to the indispensable role of this abstraction in describing an experienced phenomenon as an interaction in which the the observed system has a causal effect on the measuring instruments.

The use of the classical concepts for describing phenomena makes the description unambiguous. Thus, by definition, a phenomenal object is one which is describable by classical concepts. However, to make this description of a phenomenon as an observational interaction unambiguous, we must specify how we partition the whole interaction into observing system and observed object. Since how we do this is arbitrary, depending on what we want to describe the phenomenon as an observation of, it may appear that in choosing to describe an interaction as an observation of a *particular* phenomenal object, the scientist "creates" the properties correlated with his descriptive terms. But this is not so; the properties to which these terms refer are there in the interaction which the scientist experiences as the phenomenon to be described. The *phenomenal* object really does possess the properties which the description attributes to it. Thus the atomic system as observed in a particular phenomenon really does have a spatio-temporal locus *or* a momentum and energy. This fact is evident in Bohr's comment that "it is essential to remember ... that the *properties of atoms* are always obtained by observing their reactions under collisions or under the influence of radiation".[14] But these are the properties of atomic systems in different interactions with specific observing instruments. We cannot conclude from this fact that the atomic system in isolation possesses *either*, much less *both* the properties attributed to it in interactions in the mode of space–time coordination and interactions described in the causal mode. Bohr's caution against speaking of "creating attributes to objects by measuring processes"[15] is meant as much against idealism and phenomenalism in which the object of description is identified with an entity which comes into existence only in the subject's experience of it, as it is against the classical realist view which assumes that the theoretical parameters we use to represent the isolated system correspond to the properties possessed by an independent reality. His comment is intended in the spirit of a critical realism in which the ontologically *independent* object with its properties not only is not the same as the *phenomenal* object observed in an interaction, but also is not even to be described through the same concepts which are well-defined in reference to the observable properties through which the phenomenal object is characterized.

Nevertheless, in Bohr's public statements the ontological status of the atomic system as an independent reality remains hanging in limbo between being simply a construct of theory (which Bohr means to deny in order to speak of phenomena as interactions) and the classical conception of an independently real substance possessing properties corresponding to the observable properties used to characterize our experience of its phenomenal appearances. This fact, however, does not preclude drawing from complementarity some implications of what a complementaristic conception of such an independent reality might involve.

As we shall see in the last chapter, in rejecting instrumentalism, complementarity defends a realistic interpretation of scientific theory, but it does not do so by trying to make particle and wave pictures correspond to an independent reality. Nor does it hold that the abstract theoretical representation of the isolated system pictures the unobserved object. The real object, the atomic system to which quantum mechanical descriptions apply, enters the description through the necessity for describing observation as an interaction. It is true that in this interaction we cannot represent theoretically the state of the object system distinct from that of the measuring system, as was to be expected once the quantum postulate is accepted. The descriptive ideal stipulated by the traditional framework of determining by observation the classical state of an isolated system is consequently impossible to attain. However, this fact hardly implies that in complementarity we dispense with any reference to the nature of an independently real entity the behavior of which quantum mechanics was formulated to describe. What it does mean is that we cannot expect to describe that object as it was described in the classical framework. Thus the "generalization" of the classical conceptual framework which Bohr advocates does much more than eliminate the ambiguities which result from viewing quantum theory on the basis of the classical presuppositions. Ultimately in proposing to revise the basis for the objectivity of scientific descriptions, Bohr's revolution alters the concept of physical reality.

Although the subdivision of the observational interaction into an observing system and the observed object takes place only in the description of the phenomenon as an observation of a particular object, to avoid any misunderstanding, we should note that in classical physics originally the "observing system" was essentially the human perceiver. Thus classically the distinction we make between the observing system and the observed system *in the description* of a phenomenon would generally correspond to the ontological distinction between subject and object. But this view regards the phenomenon as the property of an experiencing subject. The properties of *subjectively experienced* phenomena are regarded as causal effects of the properties possessed by an independent physical reality, and these phenomena are the subject's consciousness of that independent reality.

But in complementarity, since Bohr regards the phenomenon as the objectively given datum for experience, the distinction between the observing system and the observed object cannot correspond to the ontological distinction between the subject which has the experience and the phenomenon which is experienced. The distinction between the observing system and observed object is made "within" the description of the phenomenon as an observational interaction. Whereas the distinction between subject and object is made "between" the experiencing subject and the phenomenon experienced. As we have seen, because the distinction between observing system and observed object is arbitrary, any given phenomenon may be described differently as an observation of different "phenomenal objects". Therefore this distinction cannot correspond to the distinction between the entities which interact to produce the experienced phenomenon and the subject that experiences the phenomenon.

Nevertheless, it would be a confusion to describe observation as an interaction, unless one presupposed a different distinction between the entities that interact to produce the experienced phenomenon described as an observation. Complementarity cannot deny this distinction and still insist that observation in atomic physics must be described as a discontinuous interaction, for what sense would it make to speak of phenomena as arising in interactions unless in some way the entities which interact can be distinguished *prior* to the interaction? If this distinction is considered to be *only* a feature of a particular way of describing phenomena, then in principle it would be possible to find a different way to describe observations such that space–time description and causal description could be combined as in the classical ideal. Since we know Bohr absolutely rejected any such search, he cannot mean the distinction between interacting systems made to describe phenomena to be merely a consequence of the quantum theoretical means of describing atomic phenomena. Phenomenalism may be able to do without the notion of an independent physical reality, but it cannot do so and at the same time hold that a phenomenon must be described as an interaction between physical systems which are *not* phenomenal objects. Precisely because complementarity is erected on the quantum postulate and because this postulate entails the description of observation as interaction, it follows that any purely phenomenalistic reading of Bohr's viewpoint cannot be compatible with its starting points. Thus although Bohr makes it very clear that wave and particle concepts do not picture the properties of an independent reality, and the quantum theoretical representation of the state of the isolated system is an "abstraction", he must presuppose the reality of the atomic systems which quantum mechanics describes as the physical systems which interact with the measuring apparatus to produce the phenomena which confirm the theory.

From these facts one might be tempted to infer that complementarity contradicts itself by holding that there must be a distinction between observed object and observing system *prior* to their interaction to produce an experienced phenomenon and that this *same* distinction can be specified only in the context of a description of that phenomenon *after* it has taken place. However, these are really two separate distinctions. On the one hand, there is the distinction which we make to describe the phenomenon unambiguously as an interaction in which the observed object has a causal effect on the observing system. This is a distinction drawn within the *description* of an interaction. But because we must characterize the interacting systems in the quantum mechanical formalism as forming an individual whole, we cannot conclude that the phenomenal object thus described is a picture of the independently real atomic system which interacts with the observing system.

On the other hand, when we describe the experienced phenomenon as arising in an interaction between the measuring instruments and the atomic systems, we presuppose a real distinction between the observing system and the system with which it comes into physical contact so as to produce the phenomenon which is described as an observation of the physical systems whose behavior is the concern of quantum theory. This entity which interacts with the observing system to produce the interaction which is described as an observation of a certain determinate object is *a* cause of the *phenomenon* experienced. But of course the observing instruments also form part of the cause of that phenomenon. However, the individuality of the interacting systems implies that any partition assumed in the description is arbitrary. Hence the same phenomenon could just as well have been described as an observation of a different phenomenal object. For this reason any description which fails to include a clear specification of how the observing system is distinguished from the phenomenal object is liable to ambiguity. It follows, then, that the two different distinctions appear at different levels in complementarity. What does not follow, of course, is that the terms which are unambiguous for describing phenomenal objects can also be used unambiguously to describe objects apart from their phenomenal manifestations.

The description of observation is not the description of some mysterious psycho-physical effect that the independently real object produces on the subject to create the phenomenal object. It is, rather, the description of an individual physical interaction partitioned for purposes of unambiguous communication into observing system and observed object. Observation *is* interaction, not the *result of* interaction. The interaction which science describes as the observation that supports a theory is the objective datum, not the subjective property of some private conscious subject. Physical science cannot express its description of phenomena in terms which would place the distinction between observed and observing systems within the subjective experience of a human consciousness, for such a distinction could not be specified unambiguously through the use of the classical mechanical concepts.

In this way, Bohr makes clear his belief that complementarity preserves the objectivity of science by eliminating subjective elements from the description of phenomena. Thus if "observer" is identified with the experiencing subject, complementarity preserves the ideal of the "detached observer", though our understanding of it is revised:

> The notion of complementarity does in no way involve a departure from our position as *detached observers* of nature, but must be regarded as the logical expression of our situation as regards objective description in this field of experience. The recognition that the interaction between the measuring tools and the physical system under investigation constitutes an integral part of the quantum phenomena has not only revealed an unsuspected limitation of the mechanical conception of nature, *as characterized by the attribution of separate properties to physical systems*, but has forced us, in the ordering of experience, to pay proper attention to the conditions of observation.[16]

As it happens, Bohr's claim that "the notion of complementarity does in no way involve a departure from our position as detached observers of nature" was challenged by his favorite pupil and friendly critic, Wolfgang Pauli. Upon reading Bohr's draft of the essay from which the above quotation was taken, Pauli replied to Bohr in his own inimitable style overladen with the sarcasm and irony for which he was both infamous and beloved:

> Under your great influence it was indeed getting more and more difficult for me to find something on which I have a different opinion than you. To a certain extent I am therefore glad, that eventually I found something: the definition and the use of the expression "detached observer". ... According to my own point of view, the degree of this "detachment" is gradually lessened in our theoretical explanation of nature and I am expecting further steps in this direction. ... it seems quite appropriate to call the conceptual description of nature in classical physics, which Einstein so emphatically wishes to retain, "the ideal of the detached observer". To put it drastically the observer has according to this ideal to disappear entirely in a discrete manner as a hidden spectator, never as actor, nature being left alone in a predetermined course of events, independent of the way in which the phenomena are observed. "Like the moon has a definite position" Einstein said to me last winter, "whether or not we look at the moon, the same must also hold for the atomic objects, as there is no sharp distinction between these and macroscopic objects. Observation cannot *create* an element of reality like a position, there must be something contained in the complete description which corresponds to the *possibility* of observing a position, already before the observation has been made." ... I consid-

> er the impredictable [sic] change of the state by a single observation—in spite of the objective character of the result of every observation and notwithstanding the statistical laws for the frequencies of repeated observation under equal conditions—to be *an abandonment of the idea of the isolation (detachment) of the observer from the course of physical events outside himself.*[17]

Though Pauli had worked with Bohr on the original Como paper and always regarded himself as an advocate of complementarity, he apparently did not distinguish Bohr's view from the interpretation which Einstein had imputed to him. Both Pauli and Einstein confuse the distinction made in the description of the phenomenon as an interaction between the observing system and the atomic object with which it interacts with the distinction between the subject that experiences the whole phenomenon and the phenomenon which that subject (or any other) experiences. Unfortunately Bohr does not seem to recognize that the common phrase "detached observer" here confuses the issue by tending to identify the experiencing subject, the "observer" in classical physics, with the "observing system", which, according to Bohr's view, is treated in the description of the phenomenon as a physical system partially causing the phenomenon given to the subject. Thus Bohr replied to Pauli's criticism:

> As always, you touch on a very central point. A phrase like "detached observer" ... used in connection with the phrase "objective description" ... had to me a very definite meaning. In all unambiguous account it is indeed a primary demand that the separation between the *observing subject* and the *objective content of communication* [*i.e.*, the phenomenon to be described] is clearly defined and agreed upon. ... this condition is indispensable in all scientific knowledge. ... It appears that what we have really learned in physics is how to eliminate subjective elements in the account of experience, and it is rather this recognition which in turn offers guidance as regards objective description in other fields of science. To my mind this situation is well described by the phrase "detached observer".[18]

In his reply to Bohr's letter, Pauli distinguished between the "ideal of objective description" which he "warmly embraced" and the "ideal of the detached observer" which he "found much too narrow" on the grounds that such an ideal would forbid any reference to the physical conditions of the observing subject.[19] Of course Bohr had already insisted that the description of a phenomenon must include a description of the whole interaction between observing system and observed object, all of which forms part of the whole phenomenon to be described by an adequate theory. Thus certainly Pauli was misunderstanding Bohr's intention in the phrase "detached observer".

In his next reply to Pauli, Bohr stood by his original choice of terms and stressed both the conceptual aspect of his claim and the fact that there was no question of "subjective interference" in his account of the description of experience in complementarity. In the process he distinguished his position from the "critical philosophy" of Kant:

> To characterize the scientific pursuit I did not know any better word than detachment. ... I wanted to stress the difficulties which ... have had to be overcome to reach the detachment required for objective description or rather for the recognition that ... we meet with no special observational problem beyond the situations of practical life to cope with which the word "observer" has been originally introduced. ... the lesson ... should teach us of the necessity for caution in looking for unambiguously communicable experience. ... I may for a moment remind of the days of so-called "classical" physics and "critical" philosophy, when in the description of the course of events the role of the tools of observation was disregarded and space–time co-ordination and causality were considered a priori categories. It is true that before the epistemological aspects of the problem were so widely cleared up, a certain confusion was prevalent, but after the thorough lesson which we have received, the whole situation including that of classical mechanics appears in a new light. ... The point I wanted to stress ... is that, just by avoiding any such reference to subjective interference which would call for a misleading comparison with the classical approach, we have within a large scope fulfilled all requirements of an objective description of experience obtainable under specified experimental conditions.[20]

While there need be no separation between observed and observing systems on a physical level—indeed there cannot be if there is to be an observation at all—through requiring a description by a "detached observer", objectivity demands that we "eliminate subjective elements in the account of experience" by regarding the whole phenomenon (described as the observational interaction between systems) as that which is given to the experiencing subject. Thus the "observer", in this sense the experiencing "subject", is detached from the phenomenon. However, within the description of that phenomenon as an observation, the observing system (as a physical system) must be represented as interacting with the observed object (the atomic system) in a way which gives the whole the individuality required by the quantum postulate. In this sense the observing system, as part of the phenomenon, does indeed create the properties ascribed to the phenomenal object.

4. The Relationship between Complementarity and Kantian Epistemology

Bohr's claim that the classical concepts are necessary for an objective description of experience may seem similar to Kant's view that the concepts of space, time, and causality can be known to apply to experienced phenomena *a priori*. Furthermore his view that these concepts apply only to phenomenal objects and cannot be used to characterize an independent physical reality seems to parallel Kant's ban on the application of these concepts to a transphenomenal reality. These facts have given rise to the view held by some of the most perceptive of Bohr's interpreters that his position contains Kantian elements supporting a subjectivistic reading of complementarity.[21] Since Bohr specifically stated complementarity provides an objective description of experience, it would seem that virtually any such reading would be contrary to his intent, though not necessarily inconsistent with his position. Before finishing with the epistemological comments of this chapter, it will help dispel the subjectivistic reading of complementarity to demonstrate in some detail that the apparently Kantian elements of complementarity are deceptive and that these two viewpoints are significantly different.

By refusing to define objectivity with respect to how experience originates, Bohr turns his back on the whole tradition of representationalistic epistemology. Although there is a superficial resemblance between some aspects of Bohr's view and a generally Kantian epistemology, it must be cautioned that often such apparent similarities are created by equivocating on such terms as "experience" and "objectivity" which change meaning in the shift from Kantian epistemology to complementarity.

Ever since Kant it has been the tendency of epistemology to ground the objectivity of knowledge based on subjectively experienced phenomena by appealing to the conceptual element entering into the knowing process. Indeed Bohr would agree that the objectivity of science is a function of the concepts with which experience is described. In this limited sense it would be possible to call Bohr a Kantian but such an attenuated definition of the label provides little enlightenment. In a much more profound sense the differences between Bohr and Kant can and should be emphasized.

It is perhaps surprising that not a single direct reference to Kant can be found in Bohr's published works. Because of the apparent similarities between complementarity and Kant's epistemology, I examined both published and unpublished papers and letters to detect positive evidence of any possible Kantian influence. Bohr's very few references to "the critical philosophy" are all negative, *contrasting* the Kantian position (at least as Bohr understood it) to the new framework. The best example is the following:

> Above all [the success of classical mechanics] deeply influenced the general philosophical attitude in the following centuries and strengthened the view that space and time as well as cause and effect had to be taken as a priori categories for the comprehension of all knowledge.
>
> The extension of physical experience in our days has, however, necessitated a radical revision of the foundation for the unambiguous use of our most elementary concepts, and has changed our attitude to the aim of physical science. Indeed, from our present standpoint physics is to be regarded not so much as the study of something a priori given, but rather as the development of methods for ordering and surveying human experience. In this respect our task must be to account for such experience in a manner independent of individual subjective judgment and therefore objective in the sense that it can be unambiguously communicated in the common human language.[22]

This passage and the letter to Pauli cited at the end of the previous section tell us that Bohr regarded the critical philosophy as an attempt to prove the universality and necessity of the Newtonian framework by arguing that its concepts refer to forms imposed *a priori* on the formation of experienced phenomena by the perceiving subject. In order to understand why Bohr regarded the Kantian position (as he understood it) as opposed to complementarity, we must notice the very important fact that Bohr based his whole argument for the correct use of classical concepts on a *physical* discovery, a purely *contingent* fact. It should be recognized at once that a true Kantian approach would never argue, as Bohr has argued, that such a physical discovery would demand a change in the proper use of the concepts which give experience its form. If Bohr were a Kantian, he could argue, as he does, that the classical concepts are indispensable, but then his claim that the quantum postulate demonstrates that the classical framework is no longer tenable would be a complete *non sequitur*. Bohr's rejection of the Kantian outlook thus follows from the fact that as he understood Kant, the critical philosophy was designed to show that the classical concepts were independent of the content of any particular experience and thus could not be shown to be inadequate by any empirical discovery. Yet complementarity was designed around an empirical discovery which has proved to be the nemesis of the classical view. Within Kant's own analysis the concepts are employed prior to the origination of experience, and therefore it would be a denial of Kant's derivation of the concepts to argue that the appearance of a new set of phenomena (the phenomena of atomic physics) demands a change in the proper use of concepts. Yet this is precisely what Bohr has argued.

Nevertheless, Bohr's restriction of the classical concepts to the description of phenomenal objects has a certain Kant-like appearance, for it issues in statements about the logical requirements for the proper use of concepts in describing nature. But such an appearance is revealed to be deceptive when we recall that these claims have nothing to do with *how* experienced phenomena arise in the subject's consciousness, as the Kantian statements do, but have to do only with communicating a *description* of a phenomenon as an objective datum already given in experience. For this reason, Bohr's talk about the use of concepts is tellingly non-Kantian.

It is *not* Bohr's intention that complementarity provides an account of how the subjectively intuited data of observation are furnished with conceptual form and thus synthesize the representations of experience. Yet this is precisely Kant's task. It *is* Bohr's intention that complementarity accounts for the limits of the unambiguous use of concepts, specifically the mechanical state parameters: space, time, energy, and momentum. For Kant two of these "concepts" are not "concepts" in his sense at all, but rather forms of intuition without which sensibility is impossible. Bohr nowhere commits himself to the position of Kant's "Transcendental Aesthetic" but claims only that reference to space and time is necessary in order to communicate unambiguously the results of an observation. By his recurrent use of the phrase "proper use of concepts", Bohr signifies the application of concepts in describing experience unambiguously in accord with the demands of objectivity. Kant's concern with the proper use of concepts is altogether different. Kant's conception of experienced representations with the objective features of the concepts synthesized into them as the product of the understanding is radically different from Bohr's notion of experience as the starting point for empirical knowledge to which the classical concepts are applied to express an objective description. These different ways of dealing with experience reflect different understandings of the task of epistemology. Confusion of the two will only bar the way to a proper understanding of Bohr's viewpoint.

The divergence between Bohr and Kant is also obvious when we consider where the concepts play their role in the acquisition of scientific knowledge. For Bohr the proper use of concepts is to make possible objective communication in the scientific description of experienced phenomena, while for Kant the concepts achieve their proper function within the understanding to give form to experienced representations. Thus two meanings of "concepts" must be distinguished. The Kantian concepts play their role prior to experience and give form to that which is experienced; "concepts" for Bohr have their role after experience has taken place and is to be described objectively. Bohr holds that the concepts of space and time have the empirical reference they do have because they refer to the phenomenal experience of space and time. Although Bohr does indeed insist that "in interpreting observation use has always to be made of theoretical notions",[23] when we remember that "interpreting observation" means describing an interaction, this claim does not imply the "idealist" view that our concepts "create" the world of experience they are used to describe. Although the concepts used to describe experience are chosen because they permit objective description, the justification for this is not to be found in a formative role of the concepts in producing experience, but rather in the fact that we have learned these concepts from experience. Because the Kantian concepts play their role prior to the formation of experience, Kant is faced with deriving the concepts from a source other than experience, namely reason itself; Bohr makes no such attempt and clearly regards the Kantian *a prioristic* derivation of the concepts as opposed to his view.

However, Bohr's insistence on retaining the classical concepts should not be read as an allegiance to the classical realist interpretation of the mechanistic framework. In complementarity, the objectivity of description is grounded in the empirical reference of the descriptive concepts to properties of the phenomenon which is described, rather than through appeal to a transphenomenal level with properties corresponding to these descriptive concepts. Bohr stresses that in resting its claim to objectivity on determining the mechanical state of an isolated system as a picture of an independent physical reality, classical realism staked its claim on a presupposition which has turned out to be empirically unsupported in the quantum domain. By placing interaction at the heart of the scientific description of observation, the quantum postulate has brought home the epistemological lesson that objectivity cannot be consistently defined through the classical appeal to the representation of an isolated system as a picture of an independent physical reality, but it can be achieved if it is defined in terms of an unambiguous description of experience. Bohr retains both the theoretical representation of an isolated system and the "detachment" of the observer from the object of description, but he revises our understanding of both. It may be possible, of course, to develop a philosophical account of science which abandons these notions, but it is not complementarity. Bohr's refusal to abandon these ideals, and his consequent refusal to abandon the classical concepts once their use is properly understood, underscores the fact that far from representing a surrender to subjectivity, complementarity is an attempt to reckon with new physical facts and yet preserve objective description.

CHAPTER EIGHT

COMPLEMENTARITY AND THE NATURE OF PHYSICAL REALITY

In the previous chapter we saw how Bohr argues that complementarity provides an objective description by retaining the classical concepts for describing phenomena but restricting them from describing an independent physical reality. As part of his defense of this view, Bohr holds that the descriptive concepts of classical physics are not *a priori* categories but are developed to communicate unambiguously a description of actually experienced phenomena. It is this fact which gives these concepts a direct reference to aspects of experience, thus providing the empirical basis on which a theory can be accepted or rejected.

However, the acceptance of a theory is not solely a function of its empirical confirmation; in this respect quantum theory is not found wanting. But, as we have seen at least in the case of Einstein, a theory may be rejected because it fails to conform to an ideal stipulating the sort of description of nature which an acceptable theory must provide. The ideals of description stipulated by a conceptual framework are in turn shaped by the the presuppositions of that framework concerning how nature can be described. Insofar as quantum theory does not conform to the descriptive ideals of the classical framework, Bohr's foundational argument seeks to show how the acceptance of quantum theory entails the generalization of that framework. The growth of science into new areas of experience, he argues, reveals that the presuppositions made by the classical framework turn out to be true only in special cases and thus must be replaced by more general assumptions. Complementarity claims to be just such a generalization of the classical framework. This generalization implies that we must revise our assumptions about how the descriptive concepts employed by the framework relate to the objects they are used to describe. This is the epistemological lesson of complementarity.

In this chapter I will argue that this epistemological lesson, and indeed the whole framework of complementarity, makes sense only on a realistic interpretation of the scientific description of nature. Bohr's method of argument must assume that in describing nature through our theories, we learn empirically from the success or failure of these theories what nature permits the scientist to presuppose about it in order to describe it unambiguously. This is simply another way of saying that science informs epistemology, or that the nature of reality molds our description of it. Furthermore, the fundamental *physical* claim of complementarity, the point at which it changes the classical presuppositions, is the assertion that to describe a phenomenon as an observation of a physical system, theory must represent that phenomenon as a physical interaction in which the interacting systems form an individual whole such that after the interaction they have changed state in an "uncontrollable" discontinuous manner. This view implies that insofar as empirical knowledge is based on the need to formulate an unambiguous description of observation, what we know about nature is the result of a physical causal process in which an independent reality produces an effect on the observing instruments, whether these instruments are human sense organs or pieces of laboratory equipment. Both of these views are hallmarks of an essentially realistic understanding of science.

From the beginning of his scientific work Bohr understood the atomic theory in a straightforward realist sense. Of course it may be that although this outlook led him to complementarity, the finished framework he offers to replace that of classical physics is incompatible with realism. Unfortunately, Bohr virtually refused to comment on the relationship between complementarity and the nature of physical reality, leaving his would-be interpreters arguing the point ever since. Thus in this chapter I will have to reconstruct what complementarity must assume about the nature of physical reality with only the barest minimum of direct statements from Bohr. For this reason, the weight of my realistic reading of complementarity falls on showing that only through such a position can complementarity be made consistent with Bohr's "epistemological lesson". We will begin by considering the realistic orientation of Bohr's attitude to his life's work, then turn to the alleged anti-realism of complementarity. Next, on the basis of Bohr's epistemological lesson, we shall consider the role of an independent physical reality in complementarity. Finally, we will conclude with a comparison of this view of how physical reality may be described with the classical view.

1. Bohr's Realism with respect to Atomism

The startling character of the innovations required by quantum mechanics may well divert our attention from the important fact that at bottom quantum theory is simply the most recent chapter in the long story of atomism. The phrases "atomic theory" and "atomic physics" which appear again and again in the titles of Bohr's essays no doubt had a freshness and dramatic quality which has been dimmed by the currently commonplace use of "atomic". We should recognize at the outset that Bohr's great overriding interest in atomic theory is the central theme of his life's work as a physicist.

The period of Bohr's youth marks a fundamental change of outlook regarding atomism. Prior to this century the atomistic approach was generally held to be purely speculative, but by 1900 the development of new experimental techniques involving radioactive phenomena and the accumulation of experimental evidence changed the expectation of physicists in the *avante-garde* of Bohr's generation. Rightly or wrongly, Bohr and many of his contemporaries not only believed that the atomistic hypothesis was empirically vindicated, but also approached atomic theory in a fundamentally realistic frame of mind. Although they may have often been seriously puzzled about what these atoms were like and they often had to defend their conviction that matter is really made up of atoms, this was a move which Bohr and his co-workers made at the beginning of their scientific careers without the need for any intellectual conversion or the exertion required to overcome anti-realist scruples about the reality of such theoretical "constructions". Just as to the religious apologist it is never God's existence which is ever really at issue, but His nature which needs defense and elaboration, so to Bohr it was never the existence of the objects of quantum mechanical description which was in question, but only how we are to understand that description.

As Bohr understood the situation, his belief in the reality of atoms was empirically justified: "We now know, it is true, that the often expressed skepticism with regard to the reality of atoms was exaggerated; for, indeed, the wonderful development of the art of experimentation has enabled us to study the effects of individual atoms."[1] The discoveries of the electron and the atomic nucleus as the presumed constituents of atoms greatly strengthened this outlook. Indeed for thinkers such as Bohr they empirically justified the atomic hypothesis, thereby focusing attention on the task of describing atomic systems by building a "model" describing their components and interrelations. Thus, as we have seen, when Bohr adopted the quantum postulate, he wrote, "I have perhaps found a little bit about the structure of the atom ... which perhaps is ... a tiny little bit of the *reality*."[2] Originally it was supposed that such models could be constructed by holding that the constituents of these atomic systems were arranged in a fashion that obeyed the laws of classical mechanics and electrodynamics. Bohr's scientific work began with the view that this supposition could no longer be upheld, but it also began with the conviction that the successful theory would provide knowledge of the behavior of real physical systems. One should not ignore the effect which work in Rutherford's laboratory must have had in convincing the young Bohr that the new phenomena which were being discovered empirically determined the structure of the atom, much less overlook the profound influence which the success of his revolutionary atomic theory of 1913 must have had on shaping the outlook of this twenty-eight year old natural philosopher. This frank acceptance of the reality of the atom as a physical system demonstrates that Bohr begins his scientific work by taking leave of that "skepticism with regard to the reality of atoms" resulting from the phenomenalism of Mach with which he has been mistakenly associated.

Throughout all of his early work from 1911–1925 Bohr repeatedly stated that the goal of atomic theory is to understand the properties of matter, specifically the properties of the different chemical elements as represented in the periodic table, through an appeal to their differing atomic structures.[3] One might imagine that perhaps this early enthusiasm for the reality of atoms waned after he embraced complementarity, but in 1929, two years *after* Como, he saw fit to begin the concluding essay of *Atomic Theory and the Description of Nature* with the following clear statement of his realistic interpretation of atomism:

> Natural phenomena, as experienced through the medium of our senses, often appear to be extremely variable and unstable. To explain this, it has been assumed, since early times, that the phenomena arise from the combined action and interplay of a large number of minute particles, the so-called atoms, which are themselves unchangeable and stable, but which, owing to their smallness, escape immediate perception. Quite apart from the fundamental question of whether we are justified in demanding visualizable pictures in fields which lie outside the reach of our senses, the atomic theory originally was of necessity of a hypothetical character; and, since it was believed that a direct insight into the world of atoms would, from the very nature of the matter, never be possible, one had to assume that the atomic theory would always retain this character. However, what has happened in so many other fields has happened also here; because of the development of observa-

> tional technique, the limit of possible observations has continually been shifted. We need only think of the insight into the structure of the universe which we have gained by the aid of the telescope and the spectroscope, or of the knowledge of the finer structure of organisms which we owe to the microscope. Similarly, the extraordinary development in the methods of experimental physics has made known to us a large number of phenomena which in a direct way inform us of the motions of atoms and of their number. We are aware even of phenomena which with certainty may be assumed to arise from the action of a single atom. However, at the same time as *every doubt regarding the reality of atoms has been removed* and as we have gained a detailed knowledge of the inner structure of atoms, we have been reminded in an instructive manner of the natural limitation of our forms of perception.[4]

This passage shows that the problem for Bohr was never the "reality" of atomic systems, which he took to be demonstrated by the empirical evidence; the fundamental question for him was always to learn how to *describe* these atomic systems in a way which accounted for "the phenomena which arise from the combined action and interplay of ... the so-called atoms". Only this position makes sense of the historical path that Bohr took to complementarity.

Because of the vital relationship between Bohr and Heisenberg during the period from 1924 through 1927 and the common tendency to interpret the young Heisenberg as an anti-realist, the view is sometimes expressed that complementarity was born in a spirit of anti-realism. Heisenberg's well-known but ill-understood pronouncement that his matrix mechanics was intended to dispense with a physical interpretation of the use of formal parameters for representing systems which are in principle unobservable has been understood as endorsing instrumentalism, thereby tending to support a similar reading of complementarity.[5] However, the historical record belies this interpretation of either Heisenberg's or Bohr's position. Matrix mechanics had rejected the attempt to form a space–time description of the atomic system, but Heisenberg clearly regarded his mathematical formalism as descriptive of the atomic system. He recalls that "The mathematical scheme had for me a magical attraction ... perhaps here could be seen the first threads of an enormous net of *deep-set* relations."[6] Both physicists regarded their theories as attempts to describe real atomic systems. The question posed by the quantum revolution was never seen as asking *whether* the concepts of quantum theory referred to an "external reality", but rather it was seen as asking *how* its concepts applied and in what circumstances they could be used. When Heisenberg and Bohr talk about the impossibility of *visualizing* atomic systems *in the terms of classical physics*, their point is that the classical descriptive concepts must be limited in their use because their definition depends on classical presuppositions which turn out to be false in the atomic domain. But neither physicist means to reject the goal of physics as providing knowledge about real physical systems which produce the phenomena we experience, as does the anti-realist. When Heisenberg refers to discovering matrix mechanics, he recalls very vividly that he felt that "*through the surface of the atomic phenomena* I was looking at a strangely beautiful interior, and felt almost giddy at the thought that now I had to probe this wealth of mathematical structures which nature had so generously spread out before me".[7]

In fact the positivist distrust of atomism no doubt produced an emotional turning away from the attitude of Ernst Mach, who epitomized the rejection of a realistic interpretation of atomic theory. Heisenberg claims that "I have never read Ernst Mach quite seriously. ... I was never much impressed by Mach. ... I would say that Mach was always a bit too formal for me. It was too—I would not say negative, but too modest in what he wanted."[8] Yet by all accounts at this time in his life Heisenberg was much closer to an anti-realist interpretation of theories than was Bohr. Thus the claim that instrumentalism had a historical role in producing complementarity rests on a misunderstanding of Heisenberg's method of arriving at the "new quantum theory".

2. The Anti-Realist Tendencies in Complementarity

Bohr's most sensitive commentators have observed a confusion in complementarity. On the one hand, Bohr's realistic view of atomic systems and the objects described by quantum theory underlies his whole interactionist argument on the nature of observation. On the other hand, anti-realist tendencies arise in combating Einstein's use of *classical* realism to defend his claim that the inability of quantum theory to describe its objects as existing in well-defined classical states betrays the incompleteness of the theory. Bohr's rejection of *classical* realism in his debate with Einstein and his restriction of the classical concepts to refer only to phenomena suggest that he means to endorse some form of anti-realism. In defending his claim for the completeness of quantum theory Bohr argues that the use of classical concepts to represent quantum mechanical objects as waves or particles existing independently of observation refers to an "abstraction" or "idealization" necessary for theoretical purposes but does not picture an independent reality. Viewed from the classical framework, this defense might well be taken as supporting a phenomenalist view of reality or an instrumentalist view of scientific theory. Indeed, this seems to be how Einstein understood him. However, as we have seen in the previous section, Bohr certainly viewed the atomic system as an independent physical reality. Thus it is not surprising that viewed from the framework of complementarity, Bohr's position entails a rejection of phenomenalism and an appeal to the nature of an independent reality which belies the instrumentalist interpretation of theories.

Bohr holds that in describing experience, the task of science is to bring order into an ever growing range of phenomena. From the definition of the state of an isolated system derived from the theoretical formalism, we make predictions about the phenomena which will be experienced by describing them as observational interactions between observing instruments and microsystems. When the theoretical predictions are confirmed by experienced phenomena, we regard the theoretical description as adequate. Although the predictions of what will be observed must be expressed in the classical concepts which have unambiguous reference to aspects of the experienced phenomena, the theoretical representation of the atomic system as isolated does not enable us to "visualize" the properties of an independent physical reality using the concepts which are well defined for describing phenomena. Because the most striking difference between the classical framework and Bohr's viewpoint appears in Bohr's strong insistence that we cannot *visualize* through wave and particle pictures the properties of the atomic system which produces the phenomena that confirm quantum theory, Bohr is led into making claims with which any instrumentalist would emphatically agree: "We meet here [*i.e.*, in quantum physics] in a new light the old truth that in our description of nature the purpose is not to disclose the real essence of the phenomena but only to track down, so far as it is possible, relations between the manifold aspects of our experience."[9]

Although comments such as this are certainly compatible with anti-realism, such a position is not consistent with the argumentation that took Bohr to complementarity. Of course he would agree that the theory is an instrument which makes possible more and more ordered descriptions of phenomena, but the success of that instrument is based on a surprising discovery learned from nature, that expressed in the quantum postulate.

Against the instrumentalistic reading of complementarity, it is clear that Bohr was not much concerned with merely the predictive devices of the theoretical formalism. His arguments are *never* based on purely mathematical reasoning from the theoretical formalism, but always are derived from what he regarded to be a fact of nature, that interaction at the atomic level takes place in discontinuous, unpredictable transitions. For the mathematically minded Heisenberg, a purely abstract formalism could perhaps be accepted as a description of real "deep-set relations in nature". In contrast, though Bohr recognized the necessity for giving up visualizable models of the atomic system, he still demanded a physical understanding of atomic processes. As Heisenberg himself reminds us, "Bohr was not a mathematically minded man, but he thought about the connection in physics."[10] Bohr repeatedly made it clear that he was searching for a conceptual understanding of how we can objectively describe the physical interactions which produce the phenomena that quantum theory treats as observations confirming its formal representations of atomic systems. Indeed, as I emphasized in my analysis of his discussions with Heisenberg, the point of disagreement between the two men was the fact that Bohr steadfastly maintained that understanding what takes place in atomic interactions is the goal of any scientific description. Successful predictions derived from the formalism must be understood as indicating that the conceptual understanding abstractly represented by the formalism is on the right track. Had Bohr been sympathetic to an anti-realistic interpretation of theory, he certainly would have regarded matrix mechanics as essentially ending the development which his model had started. Though this was in fact the reaction of some scientists, it was not Bohr's. As we have seen, he immediately pressed harder than ever to understand this strange behavior of these physical systems which had sometimes to be represented as waves and sometimes as particles.

If Bohr was indeed an anti-realist, it is strange that in spite of the attempts of instrumentalist philosophers of science to claim him as one of their own, he never made any public declaration for their side of this historical debate. Although it may seem odd to philosophers, I suspect that Bohr simply never saw the issue between realists and anti-realists as providing two tempting options. In spite of his lifelong concern with the use of concepts in the description of nature, Bohr's profession as a physicist probably blinded him to the possibility that the entire atomic domain which he and his colleagues were attempting to describe by formulating quantum theory was but a construct of their theories. No doubt he was to some extent intellectually aware of this possibility, but in spite of ample opportunity to declare for the instrumentalist option, the fact that he never did so suggests that this possibility was simply not a "living option" on his intellectual horizon. We must not forget that although he was personally very close to Denmark's leading philosopher, Høffding, in general Bohr "felt that philosophers were very odd people who really were lost".[11] Furthermore, although the period of Bohr's creative work marked a vital era in the philosophy of science, he was continually put off by what he regarded as the failure of philosophers to understand complementarity.[12] These considerations suggest that while philosophers may regard the realist or anti-realist implications of quantum mechanics as of considerable significance, we cannot presume that Bohr saw these issues in the same way.

Nevertheless, Bohr's insensitivity to the issue of the nature of physical reality is a serious liability for complementarity, for it is possible that contrary to his intentions, he backed himself into an anti-realist corner in order to escape Einstein's criticisms, thereby undermining the realism on which he based the interactionist core of his argument. If this is so, then complementarity is incoherent. Thus the question is of crucial importance.

Part of the strength of complementarity is that Bohr never falls into the mistake of trying to use phenomena as a direct bridge between scientific theory and the structure of an independent reality. Yet he avoids committing himself to phenomenalism. It was his conviction that the chemical atoms were real entities that led to his first theorizing, and the success of that theorizing certainly tended to make phenomenalist doubts about the reality of a world not observed but merely inferred from its effects seem simply irrelevant. Nevertheless, although committed to the reality of atoms, Bohr was simultaneously aware that whatever the nature of the atomic domain might be, it was not capable of being visualized by classical models. Bohr's keen sensitivity to the half-hidden nature of the atomic realm was balanced by his exceptional feeling for the experimental situations which make possible knowledge of this realm. Heisenberg recalls being amazed at this outlook: "One of the things that impressed me most was this kind of intuition of Bohr. ... he had not proved anything mathematically ... he just knew that this was more or less the connection. ... The intensity of imagination that he had—that made an enormous impression."[13]

The instrumentalist interpretation of complementarity would seem to be supported by Bohr's view that the theoretical representation of the isolated physical system is an abstraction. Bohr would agree that were we to attempt to derive from the quantum theoretical definition for the state of an isolated system, a representation of a concrete real object, we would be misunderstanding the nature of a construct of theory, needed solely for its instrumental value in allowing us to predict the outcome of future observational interactions. However, such a misunderstanding is natural enough, because this is precisely what the classical framework held. The classical claim that the state of an isolated system provided objective knowledge of the world behind the phenomena was based on the fact that it enabled one to visualize physical systems not interacting with an observer. Although the classical framework consequently regarded "visualizable" pictures as the necessary condition for an objective understanding of physical processes, once we reject this definition of "objectivity" in terms of such an alleged picture of the properties of an independent reality, we need not regard the loss of any possibility for such a classical realistic interpretation of the state of the isolated system as an abandonment of realism. For Bohr the atomic system as an independent reality produces the interaction we experience as an "atomic phenomenon"; the limitation of the classical concepts imposed by the quantum postulate only implies we cannot "visualize" this process.

To the extent that Bohr argues that the reason we must use the abstractions of physical systems as isolated particles or waves is *not* because of some alleged correspondence between such theoretical constructs and the properties of an independent reality, his position *is* in essential agreement with instrumentalism. Consequently on the assumptions of the classical framework it would be correct to infer that complementarity rejects a realistic interpretation of scientific theory. Thus, since we cannot derive a visualizable picture of the atomic system from the quantum theoretical representation of the state of the isolated system, were we to remain within the classical framework, quantum theory would force us to break the connection between theory and the presumed independently real object which (in a classical realist interpretation of science) justifies regarding the theory as providing an objective description of nature. If Bohr is to defend his claim that complementarity provides an objective description of nature, the criterion on which *classical* realism based its claim to objectivity, the "visualization" of the state of an isolated system, must be replaced. By redefining "objectivity" in a scientific description, complementarity implies a new standard for how to describe an independent reality, but it does not thereby abandon all reference to such a reality.

If we accept the ontological view of the phenomenalist–instrumentalist, statements with truth-value can refer only to phenomenal objects. "Nature" means phenomenal nature and the question of accounting for a transphenomenal cause or "grounds" of experienced phenomena is dismissed as scientifically meaningless if it refers to anything other than physiological processes or their psychological correlates. Of course complementarity holds that phenomena which are treated as observations of atomic systems must be described as discontinuous interactions between observing systems and atomic systems. The observing system is given as a phenomenal object, but the atomic system cannot be described by classical terms so it cannot be considered a phenomenal object. Thus on the phenomenalist assumption, it must be considered a construct of the theory. Consequently, a phenomenalistic interpretation of complementarity would have to assert that observation is an interaction between a theoretical construct and a phenomenal object. Such a claim seems patently absurd. Moreover, since the phenomenalist must regard the observing system as a phenomenal object but cannot regard the atomic system as a phenomenal object, these different "physical systems" must belong to distinct ontological orders. At this point complementarity must take its leave of phenomenalism, because it holds that in an observation the individuality of an interaction (implied by the quantum postulate) forces us to regard the distinction between observing system and observed object as an arbitrary one invoked for the purpose of objective description. Since this distinction may be shifted, a single interaction may be described as an observation of different phenomenal objects. Thus the concept of a "phenomenal object" has determinate meaning only within the context of a particular description of a phenomenon. For this reason the phenomenal object cannot be the same as the "system" with which the measuring instruments interact to produce the phenomenon described as an observational interaction.

In replying to Einstein, Bohr made it clear that complementarity holds that the classical concepts can refer unambiguously only to properties of phenomenal objects described in the account of a particular observational interaction; we cannot use these terms to attribute reality to properties possessed by an independently existing atomic object. This position may seem to endorse the view that the *atomic system* described by the quantum formalism exists only in the observing interaction. Such a position, sometimes called "microphenomenalism", would amount to holding realism with respect to the objects which can be described through the "everyday concepts" of the classical framework. At the same time, the microphenomenalist denies any independent ontological status to the "objects" of quantum mechanical descriptions, maintaining that these have existence only at the level of phenomena "experienced" in the observational interactions of atomic physics.[14] This view may seem to be the same as Bohr's position that the *phenomenal* object which is described in a particular observational interaction exists *qua* phenomenal object (with properties corresponding to some, but not all, of the classical state parameters) only in the interaction. Ironically, this interpretation would bring atomism around full circle to an explicit denial of the ancient Democritean thesis that nothing is real except atoms and the void. On this position, the "everyday world" of common experience is the only "reality"; what is described by the quantum mechanical formalism is only the phenomenal behavior of certain measuring instruments. Had Bohr embraced such a position at mid-life, it would have been a complete turnabout with respect to his earlier strong commitment to the reality of atoms as the constituents of material bodies. Adopting the microphenomenalist view would introduce a remarkable unnoticed and unexplained shift in Bohr's intellectual history. One cannot rule out this understanding of complementarity on such grounds alone, for there is no necessity that Bohr's intellectual path could not have taken him to a point inconsistent with his original outlook. However, such an incoherence, and its apparently unnoticed status, may well justify our distrust of this reading of complementarity.

Be that as it may, there remains a serious logical problem with the microphenomenalist reading of complementarity. As we have had occasion to observe, if we accept the quantum postulate, we must regard classical descriptions of "macro-objects" as idealizations which have been successful merely because the dimensions involved in the interactions when we observe these objects are so huge in comparison with those of the atomic domain that the discontinuity in change of state can be disregarded. Thus complementarity explicitly rules out any interpretation of *even* the classical descriptions of macro-objects in the *classical realist sense* of a one-to-one correspondence between the concepts used to describe a phenomenal object and the presumed properties of an independent reality. Furthermore, in emphasizing the arbitrariness involved in describing an observational interaction by subdividing it into observing system and observed object, Bohr makes the distinction between the observing instruments and the phenomenal object an *arbitrary* one invoked for the purpose of description. Since the description of the macro-object serving as the observing system, on the microphenomenalist view, would have to be interpreted realistically, while the description of the atomic object would refer only to a construct of the theory, the fact that this distinction may be redrawn would imply that simply as a result of a changed description "real" objects might become theoretical "constructs" and *vice versa*. The fact that according to complementarity the distinction between observing system and the observed object is invoked for the sake of description means that at the same time this distinction cannot also serve to mark an inviolable line between that which is real and that which is a construction. Hence, in order to be consistent, a "phenomenalist" reading of complementarity would have to be an "all or nothing" interpretation. Either Bohr means to hold that the task of describing nature is restricted to predicting phenomena, and the ontological nature of everyday objects is just as much hidden from us as is the real nature of possible atomic systems, or he does not embrace phenomenalism at all. The fact that Bohr holds that the "partition" between observing system and observed object is arbitrary precludes understanding complementarity in terms of a levels ontology with some objects of description being real and others constructs of our theories.

At least three arguments can be discovered to show the incompatibility between complementarity and phenomenalism. First, Bohr's conviction that it is futile to search for an alternative atomic physics denying the quantum postulate shows that he cannot have intended complementarity as a defense of phenomenalism. For the phenomenalist, since the quantum postulate refers to atomic systems, it cannot be a consequence of any fact about reality, for reality consists of phenomena and atomic systems cannot be phenomenal objects. It must, therefore, be merely a postulate of the theory, part of a particular intellectual construction for describing certain phenomena. In this case it would be completely consistent to search for a different construction which would allow a description of the same phenomena without appeal to the presupposition that observations involve a discontinuous interaction between systems. Bohr's strong belief that such an alternative was impossible thus belies the phenomenalist interpretation of Bohr's intentions.

The phenomenalist must argue that we merely choose to describe observations of atomic systems as discontinuous interactions by choosing the particular theories through which we describe such phenomena. In this case the whole interactionist analysis of observation is itself but a construct of our theories and need not be assumed to refer to any real process which takes place in a world "behind the phenomena". The description of observation as a discontinuous interaction must be, then, merely a consequence of describing observation within a theory which adopts the quantum postulate. No doubt this is exactly what Bohr holds, but the argument that this is *just* a construct of the theory assumes that we have the liberty to describe phenomena in another theory which does not accept the quantum postulate. As we know this is most definitely a view Bohr denied. We are *forced* to adopt the quantum postulate by the way the world is. Nature teaches us through experience how it may be adequately described. This is the unsuspected empirical discovery that allows atomic physics to teach an epistemological lesson about the description of nature. Bohr's view that the extension of experience into new domains *forces* the alteration of older conceptual frameworks appears in the very first paragraph of his first collection of philosophical essays:

> Only by experience itself do we come to recognize those laws which grant us a comprehensive view of the diversity of phenomena. As our knowledge becomes wider, we must always be prepared, therefore, to expect alterations in the point of view best suited for the ordering of our experience. ... The great extension of our experience in recent years has brought to light the insufficiency of our simple mechanical conceptions and, as a consequence, has shaken the foundation on which the customary interpretation was based, thus throwing new light on old philosophical problems.[15]

This attitude reveals a fundamentally realistic interpretation of the scientific description of phenomena. Bohr sounds this same note again and again throughout his work. Science demands objectivity, and objectivity demands unambiguous communicability. The need for a conceptual framework which will permit such unambiguous descriptions of experience makes the concepts we use answerable to the world we experience. Thus the objectivity of science precludes viewing conceptual frameworks as *freely* adopted presuppositions which science imposes on experience. As an empirical fact, we have discovered that the framework which is based on describing experienced phenomena as originating in an uncontrollable interaction is the framework we must adopt to describe quantum phenomena. It might be possible to *describe* a phenomenon as arising in a physical interaction between systems in which these systems loose their separate individuality. But for the phenomenalist, statements about an interaction between atomic systems and observing instruments as giving rise to certain phenomena, cannot be said to have been learned from nature, because no matter how we might characterize such an atomic object, the fact that it is *not* a phenomenal object implies that it is not part of "nature" but must be a construct of our theories. But Bohr insists that the quantum postulate expresses an *empirical* discovery, something we have learned from nature.

The epistemological lesson of complementarity is a lesson concerning how we arrive at the frameworks we accept at any stage in science:

> The lesson of atomic physics has been that we are not simply coordinating experience arranged in given general categories for human thinking, as one might have liked to say in the expressions of physical

> philosophy [presumably Bohr means the "critical philosophy" of Kant], but we have learned that our task is to develop human concepts, to find a way of speaking which is suited to bringing order into new experience and, so to say, being able to put questions to nature in a manner in which we can get some help with an answer.[16]

The realistic basis of Bohr's epistemological lesson also appears as the background for his claim that complementarity is a rational generalization of the classical framework. Were Bohr an instrumentalist, then it would be perfectly consistent to maintain that the instrumental adequacy of classical physics in its domain of phenomena means that the classical framework is perfectly acceptable in understanding the description of these phenomena. A different framework is needed for quantum phenomena, and is justified, of course, by its instrumental value in describing these phenomena. If theories are nothing but instruments, different tools for different jobs, then there is no need for a *revision* of the classical framework, we need only *limit* its applicability to "ordinary" phenomena and devise a different tool to handle quantum phenomena. But Bohr envisioned complementarity as *replacing* the classical framework.

It is precisely because Bohr did accept a realistic understanding of atomic physics that he found the quantum mechanical description of atomic processes as understood within the classical framework to be problematic and cause for revising that framework. The whole problem of wave–particle dualism arises only when we interpret the quantum mechanical description of atomic systems along the lines of *classical realism*. One could simply eliminate the paradoxical nature of this dualism by holding that the sole purpose of theory is to make successful predictions of phenomena observed under given conditions. Such an alternative was completely unacceptable to Bohr who was thoroughly convinced by his whole life's work that the atomistic conception of material bodies was neither merely a heuristic "inference ticket" nor a strictly hypothetical construct, but rather a highly confirmed theoretical description of the real structure of the natural world. Bohr could have totally avoided all of the torment which he experienced in grappling with the wave–particle dilemma by simply adopting an instrumentalist interpretation of quantum theory. Indeed, many have assumed that this is just what he did. But the history which we have followed tells a different story. Given his conviction that the new framework must revise the classical account of descriptive concepts, it was only natural that instead he should elect to redefine the relationship between the concepts used in a theory and the phenomena which confirm that theory. Had he simply adopted instrumentalism as a way out of wave–particle dualism, there would have been no need to *revise* the classical framework. He could have kept the classical framework and merely discarded the realistic interpretation of the description of nature within that framework. But in fact he did just the opposite; he discarded the classical framework and kept a realistic understanding of the scientific description of nature. What he rejects is not realism, but the classical version of it.

A second argument against the phenomenalist reading of complementarity is based on a point made in the previous chapter. The phenomenalist reaches his position on the purely epistemological grounds that since the whole reason for accepting a scientific theory lies in its ability to describe observed phenomena, the only reality which science presupposes in its description of nature is the world of phenomena. The concept of a reality behind the phenomena plays no role in scientific theory. From a scientific view, statements about such a reality are effectively meaningless.

Bohr of course does agree that such a reality cannot be pictured and that the classical descriptive concepts which must be employed to describe phenomena are of necessity banned from reference to such an independent reality. These are certainly views with which a phenomenalist would agree. But Bohr does not reach them through an epistemological analysis revealing the impossibility of knowledge about the relationship between observed phenomena and an independent reality which serves as their grounds. As we have seen Bohr is very clear that the reason he seeks to revise the classical realist view is the result of a purely *empirical* discovery, that expressed by the quantum postulate. Had there never been the need to describe interaction as involving a discontinuous change of state, Bohr would have been content with the classical realist view that the state of an isolated system pictures an independently existing reality.

A third line of reasoning also reveals the incompatibility between complementarity and phenomenalism. When Bohr speaks of "complementary phenomena" or "complementary descriptions" or "complementary information" extracted from different experimental observations, the question arises, in what sense are these different phenomenal descriptions complementary? As early as 1938, his answer clearly indicates that they are "complementary" because they are both about "the same object":

> Information regarding the behavior of an *atomic object* obtained under definite experimental conditions may, however, according to a terminology often used in atomic physics, be adequately characterized as *complementary* to any information about *the same object* obtained by

> some other experimental arrangement excluding the fulfillment of the first conditions. Although such kinds of information cannot be combined into a single picture by means of ordinary concepts [*i.e.*, by assuming that the descriptive terms in the complementary descriptions refer to "the same object"], they represent indeed equally essential aspects of any knowledge of *the object* in question which can be obtained from this domain.[17]

Bohr's position in this regard is hardly confined to a single statement; statements of this sort occur in virtually all of his later essays. Even as late as 1957, when one might believe that the tendency to support phenomenalism was at its peak, he comments:

> Experience obtained by different experimental arrangements which had to be described by different physical concepts [*i.e.*, wave and particle pictures] exhibit a relationship which we term complementary, in the sense that each of them contains information relating to *the object* and that they together and only together exhaust the definable information we can obtain about the object.[18]

What is the "atomic object" or "the same object" to which Bohr is referring? Clearly it cannot be the phenomenal object, for we have seen that he insists that different experimental arrangements require different descriptions in which the observed object is described as a particle in some and as a wave in others because the observing arrangements are part of the whole phenomenon described. Thus there can be no possibility that the object which is "the same object" in the two pieces of complementary "information" could be the *phenomenal* object. The only alternative is to assume that Bohr intends "the atomic object" which is "the same" throughout both experimental interactions to be the object which causes these complementary phenomena. One cannot speak of "the same *phenomenal* object" as appearing differently in different experimental interactions, for the phenomenal object *is* its appearance. Thus "the same object" about which the complementary phenomena are informative must be be regarded as that object which interacts with the observing instruments to cause the phenomena.

If reference to any object other than the phenomenal object is outlawed, we can have no way of describing different phenomena as revealing different observations of the *same* object. Classically, descriptions of different phenomenal appearances could be linked through the theoretical representation of the isolated system between its phenomenal appearances in observational interactions. However quantum theory must conclude that this "representation" refers only to an abstraction, not a picture of the properties possessed by the system when not observed.

Consequently although Bohr avoids talk of the *nature* of an independent reality, his talk of different phenomena as complementary presupposes the *existence* of the atomic system as an independent reality about which these phenomena provide complementary evidence. In this way complementarity can maintain that descriptions of different phenomena can be combined in a complementary fashion because the different phenomenal objects thus described are different phenomenal appearances of the *same* object. But if it holds this view, it cannot reject the view that the scientific description of phenomena requires postulating the existence of an independently real object as cause of those phenomena. However, since it has already concluded that classical descriptive concepts refer only to phenomena, then it must develop an alternative way to express the knowledge contained in quantum theory about such an independent reality. And that way must be through combining the different descriptions of distinct phenomena in a complementary fashion as implied by the consequences of the theoretical formalism. The formalism dictates how to combine the descriptions of these phenomena as descriptions of different manifestations of the same object through the theoretical abstraction of the state of the isolated system, the purpose of which is *not* to picture the properties of an independent reality (as was classically supposed) but to allow just such a complementary combination of different descriptions in order to exhaust all that can be known about the object which produces these phenomena.

If Bohr was in fact a phenomenalist, the need for complementary descriptions would have been swept from under him, for what object could he be referring to when he asserts that two complementary descriptions, each referring to different phenomenal objects, are informative about "the same object"? If we are determined not to refer to a reality behind the phenomena, then we would simply speak of different phenomena as requiring different sorts of descriptions using different concepts. In this case the *complementarity* between different descriptions would never arise.

From these arguments it appears that Bohr could not have embraced phenomenalism and yet argued for complementarity the way he did. Yet the phenomenalist tendency in complementarity is revealing, for it indicates Bohr's rejection of *classical* realism. Since Bohr rejects the classical realist description of an independent physical reality, but he cannot dispense with reference to "the atomic object" as such a reality, what complementarity needs is a revised understanding of the role of an "independent physical reality" in the interpretation of the physicists' description of the atomic domain.

Nevertheless, the idea of an independent physical reality as the grounds for observed phenomena must remain a purely hypothetical posit for the natural scientist as for the philosopher. One must not be misled into thinking that Bohr's distinction between the mode of space–time co-ordination and the mode of applying "the claim of causality" is intended to suggest that the former describes phenomena and the latter describes their "cause". It is true that we can use the causal mode to *describe* a phenomenon as an observational interaction in which the phenomenal object has a causal effect on the state of the observing system. But using the classical concepts of momentum and energy and the associated conservation principles to describe the atomic object as the *cause* of the experienced phenomenon is impossible, for these concepts are well-defined only when they refer to the phenomenal object which is distinguished from the observing system *within* a description of the whole phenomenon.

This misinterpretation of Bohr's intention arises from equivocating on the concept of "causality". Bohr's use of the term refers to the relationship between observed object and observing system as descriptive categories within the whole interaction phenomenon. The other use, which Bohr does not intend, refers to the relationship between an independent reality and its phenomenal manifestations. To regard the causal mode of description in science as providing an exit from the phenomenal realm and describing a concrete object outside of experience as the "cause" of phenomena is to mislead epistemology into the pitfalls of representationalism and the notion of a "real" object possessing at least some properties corresponding to the properties of phenomenal objects. Any mode of description employing the classical concepts can describe only phenomenal objects. If we are to describe any *concrete* object, both modes, restricted to the use of classical concepts, must describe only phenomenal objects.

Because these modes are restricted to describing phenomena, the temptation to find a form of phenomenalism in complementarity is a strong one and not without warrant. It *appears* that Bohr's insistence on the classical concepts for describing experience accords them some privileged role which may well be taken as endowing them with *a priori* rather than empirical foundations. It *appears* that Bohr's insistence that these concepts are restricted to describing phenomena, conjoined with his view that we *must* use these concepts, constitutes a ban on any reference to an independent reality. It *appears* that Bohr's reply to Einstein that it is ambiguous to use the classical concepts to accord an element of reality to corresponding properties as possessed by an entity independently of the circumstances of its observation is the same as the phenomenalist claim that such a use is meaningless. It *appears* that his reply denies that physics has any concern with an independent physical reality that produces experienced phenomena and that, therefore theories cannot be interpreted realistically.

But all of these appearances are deceiving and, if we have resolved to make complementarity into a consistent framework, may well be avoided. They all rely on two unexpressed assumptions: 1) The content of any knowledge about a reality which lies behind experienced phenomena must be communicated as the description of the *properties possessed* by an independently existing substance. 2) These properties are essentially the same as some of the properties used to describe phenomenal objects, specifically those to which the classical mechanical state parameters refer. If we deny both of these assumptions, as is logically possible as well as necessary in order to create a new understanding of physical reality at the atomic level, then all of Bohr's statements appearing to endorse phenomenalism become consistent with his need to posit the existence of an independent physical reality.

3. The Epistemic Status of an Independent Physical Reality

The previous section established that Bohr did not intend to endorse phenomenalism, and that had he done so without being fully aware of what he was doing, he would have undermined the need for complementary descriptions. Unfortunately, Bohr never makes clear in what sense we can have knowledge of the reality which causes our experiences. In this section, therefore, I must justify how we can say that we possess a theory which is informative about the entities of the atomic domain as they exist apart from our observations of them, though of course it does not "represent" or "picture" that world in the classical fashion.

Purging the appearance of phenomenalism from complementarity entailed that having knowledge of an independent reality is not in any sense akin to having phenomenal knowledge. For this reason, the classical ideal of describing the properties of an independent reality by determining the mechanical state of an isolated system cannot be realized. In both classical and complementaristic frameworks, the *phenomenal* object is described in the manner of a thing having a certain set of properties. But in complementarity the ontological status of such properties is regarded as necessarily dependent upon an observational interaction, since that which is described by terms referring to such properties is, by definition, the phenomenal object. The properties of phenomenal objects, of course, cannot be regarded as the "causes" of observational interactions, because they refer to an object which is defined only in the context of a particular description of a phenomenon. When the phenomenal object is described, various properties are attributed to it to define its state as a wave or a particle. But the phenomenal object thus described becomes a determinate object of description only in the context of a precisely specified partition between observing system and observed object in the account of the whole phenomenon. For this reason the classical descriptive concepts can refer only to the properties of phenomenal objects and not those of an independent physical reality.

Consequently, the "objectivity" secured by the use of the unambiguous classical terms to describe phenomena is a property of *descriptions*, not some order of being lying beneath the phenomenal world. Complementarity erects its defense of its objectivity on this claim, rather than on a dogmatic adherence to a traditional ontology of substances and their possessed properties. For this reason, the question of the existence of an "external world" is irrelevant to objectivity and so never is considered explicitly by Bohr. Nevertheless, the belief that scientific theories are informative about an independently existing physical reality which produces experienced phenomena is a *presupposition* of natural science that Bohr hardly seeks to challenge. It is this presupposition that gives reason to the attempt to frame ever more adequate frameworks for the objective description of phenomena.

Although Bohr refused to be concerned with problems involving the *existence* of an independent physical reality behind experienced phenomena, he was very much concerned with restricting what we could *say* about it. It may be that the natural scientist can hardy help but to assume that there is a reality existing behind experienced phenomena, but he can very much use discretion and care in what is said about it. Indeed, learning the conditions necessary for the unambiguous use of concepts is an essential way progress in science takes place. From the viewpoint of complementarity, classical realism erred, not because it *posited* an independent reality, but rather because it held that the predictive success of classical theories legitimized describing that independent reality as composed of entities possessing properties corresponding to the parameters in terms of which the theory represented the state of an isolated system.

It is both impossible and unnecessary to expect that natural science justifies its realistic outlook by empirically demonstrating that there is a reality existing independently of experienced phenomena. Although science must insist on the logical consistency and comprehensiveness of its theoretical descriptions of phenomena, it is *impossible* to demonstrate the existence of such an independent reality, for the logical consistency of a theoretical description implies only the *possibility* of this realm's existence. From the viewpoint of complementarity, such a demonstration would also be unnecessary, for the very existence of science, as seen within the framework, implies that such a reality must exist in order to be able to describe observation as interaction. Thus science must presuppose a grounds of phenomena outside the phenomenal realm before it even begins its description of observed phenomena. To ask that science empirically demonstrates that such a reality exists, would be to ask it to prove what, according to Bohr's viewpoint, it must assume.

Furthermore, as we have seen, complementarity requires positing an object as grounds of the phenomena, for it speaks of the descriptions of phenomena observed in different experimental situations as providing complementary information about the same object. Such a set of phenomenal descriptions cannot describe a single *phenomenal* object manifested in different appearances, for to speak of a phenomenal object as appearing differently in different situations is inconsistent with the definition of phenomenon. Nevertheless, Bohr does argue that objective knowledge requires relating these different descriptions as providing a complete account of the behavior of the same object. Thus Bohr implicitly invokes the notion of an independent reality, for only through this conception can complementarity justify its claim that different phenomenal descriptions must *complement* each other to provide a complete description of the behavior of atomic objects.

The phenomenalist interpretation of complementarity misses the point that what is most revolutionary about this framework is the suggestion that in the combination of complementary descriptions of phenomenal objects, we convey information about an independent physical reality. *The very fact that it is possible, within the context of quantum theory, to combine these descriptions in a complementary fashion tells us something about the object which is the grounds of these phenomena.* Though this conclusion seems necessitated by Bohr's whole argument, it has not been generally recognized, because we are so accustomed to expecting that scientific knowledge is representational (as indeed it is for the phenomenal object), that when we discover the impossibility of using terms which are well-defined for referring to phenomenal objects to describe the *cause* of these phenomena, we conclude that any knowledge of this cause is impossible. However, there is no necessity for accepting the assumption that "to have knowledge of an object" means to be able to form a representation of that object as a substance possessing properties. Indeed, if nature is to be described within the framework of complementarity, then there is every reason for rejecting such an assumption.

Thus complementarity stipulates that within this new framework, "to have knowledge of the object which is the grounds of complementary phenomena" means to be able to formulate theoretical structures which permit different phenomenal descriptions to be regarded as complementary descriptions of the same object. *These very structures express the cognitive content of that theory concerning the atomic system which is the independently real object which grounds the phenomena.* It is true that such structures do not permit forming a *representation* of that object. Indeed, they forbid it, but it hardly follows from this fact that we are wholly ignorant concerning such an independent reality. Surely when we advance from having merely descriptions of different phenomena to a theory which allows us to coordinate these descriptions of separate phenomenal objects as descriptions of *complementary phenomenal appearances of the same object*, as well as predict statistically the probability of future phenomenal appearances, we have advanced our knowledge a great deal. What we have gained in such an advance is what we have learned about the objects which are the grounds of these complementary phenomena.

One might object that the theoretical structure which allows for co-ordinating and predicting different phenomenal appearances is just as much an abstraction as the classical notion of the state of an isolated system. We have no more warrant, one might argue, for maintaining that such a theory is informative about an independent reality than we have for so regarding the classical mechanical representation of the state of an isolated system. However, there remains a significant difference between the two viewpoints. Representing a system as isolated is *physically incompatible* with representing that system as in an observational interaction. Thus if the state of the system changes discontinuously in an interaction, as the quantum postulate maintains, it is impossible to regard the state of the isolated system as representing an independent reality. However, this impossibility does not imply that a logical bar is lowered against positing an independently real object in a theory which allows for coordinating the different phenomenal appearances such an object might present.

The key to understanding the rationale for positing an independently real object in the complementaristic description of phenomena must be found in the description of observation as interaction. Since observation must be so described, we know that the *ontological* distinction between the interacting systems is prior to observation. However, the discontinuous change of state in our description of the interaction forces us to regard as an abstraction any attempt to *represent* theoretically the system isolated from interaction. But this fact does not imply the impossibility of holding that a theory which treats different phenomenal descriptions as complementary expresses our knowledge of that with which the observing system interacts in an observational interaction. Of course it may be readily admitted that such a theoretical understanding is expressed in terms of "abstract" theoretical structures, if by "abstract" we mean being defined by terms that have no empirical reference, but such a theoretical structure need not, therefore, be regarded as uninformative. Thus while statements concerning the reality which causes experienced phenomena do not provide "empirical knowledge" in the sense that they can be confirmed directly by experience, such claims have the same epistemic status as any scientific hypothesis in that they result from a theory which is highly corroborated by contingent observations of phenomenal objects.

Specifying the formal relationships between complementary phenomenal descriptions by means of an empirically confirmed theory essentially stipulates the definition of the term "the same" through which we indicate the ontological identity of the independently real object throughout its different phenomenal manifestations. The elaboration of such relationships is the task of scientific theorizing, a task which each of the specific sciences accomplishes in different ways in order to deal with different types of phenomena. This task is often difficult and like all science may be progressively improved upon, but it is not impossible or meaningless.

Although the argument of the previous paragraphs is never presented directly by Bohr, in the following passage the fact that "atomic objects" can only mean objects as the grounds of the phenomenal appearances implies that the conclusion of this argument is one which Bohr must hold:

> In order to characterize the relation between phenomena observed under different experimental conditions, one has introduced the term complementarity to emphasize that *such phenomena together exhaust all definable information about the atomic objects*. Far from containing any arbitrary renunciation of customary physical explanation the notion of complementarity refers directly to our position as observers in a domain of experience where unambiguous application of the con-

cepts used in the description of phenomena depends essentially on the conditions of observation. By a mathematical generalization of the conceptual framework of classical physics, it has been possible to develop a formalism which leaves room for the logical incorporation of the quantum of action. This so-called quantum mechanics aims directly at the formulation of statistical regularities pertaining to evidence gained under well-defined conditions. The completeness in principle of this description is due to the retention of classical mechanical ideas to an extent including any definable variation of the experimental conditions.[19]

The distinction between phenomena and atomic objects preserved by Bohr indicates that he regards the phenomena as providing information about the independently real atomic objects. Of course we can have no means of empirically confirming such knowledge claims, other than through predictions of what phenomena will be observed. The theoretical formalism can only communicate unambiguously how the atomic object, when encountered in specific observing circumstances, will appear. To demand more of any claim to knowledge of the nature of reality behind the phenomena is hardly consistent with a philosophical account of empirical knowledge.

Although "representation" is a misleading term for referring to this knowledge of an independent reality, the term "symbol" is appropriate. A symbol *may* be representational (as in some uses of hieroglyphs or pictographs), but it need not be. In this latter sense, the equation $x^2 + y^2 = k$ in the context of analytic geometry becomes the symbol for a circle. Here a symbol expresses a functional relationship. In the context of quantum theory, the symbolic structures of the theory allow relating various descriptions of different observational interactions as providing complementary information about the same atomic object. In the following passage we should note that when Bohr speaks of "symbolic procedures" he cannot be referring to phenomenal objects, for the symbolic procedures of the formalism define the quantum state of the physical system in terms which do not have direct empirical reference, as must any description of phenomenal objects.

... we are dealing here [*i.e.*, in quantum theory] with a purely *symbolic procedure* the unambiguous interpretation of which in the last resort requires reference to a complete experimental arrangement. Disregard of this point has sometimes led to confusion, and in particular the use of phrases like "disturbance of the phenomena by observation" or "creation of physical attributes by measurement" is hardly compatible with common language and practical definition.[20]

In quantum theory such symbolic procedures are expressed in formal mathematical terms using the theoretical parameters of the quantum numbers. But since these terms have no empirical reference, they cannot be taken as referring to the properties of phenomenal objects. Thus the information which is contained within these symbolic procedures must be informative about the object as cause of the phenomena which we describe by predictions from the theoretical formalism. In this use, of course, Bohr means "phenomena" to refer to the whole interaction between atomic objects and observing agencies. Nevertheless, it is also clear that these same symbolic procedures do not provide a *representation* or "picture" of the independently real atomic object.

Bohr's reluctance to comment on the nature of physical reality in the quantum domain means that any account of the complementaristic analysis of knowledge of a reality behind the phenomena must be reconstructed from suggestions such as that indicated in the previously quoted passage. Therefore it is of interest that in his correspondence with his long-time friend, Max Born, he reveals that the position outlined above is one with which he could agree. As one who had contributed essentially to the development of the quantum mechanical formalism, Born provides an interpretation of the theory which is important in its own right. But what is of interest here, is that Born advocates essentially the sort of realist defense of quantum theory which must be the logical consequence of holding that different phenomena provide complementary evidence about the same object. Arguing explicitly that modern physics has developed a framework which of necessity goes beyond the Kantian restriction of empirical knowledge to phenomena, Born counters phenomenalism by arguing that the mathematical symbols of the theory express the physicist's knowledge of "hidden structures" which he identifies with "the objective reality lying behind the subjective phenomena":

> These are structures of pure thinking. The transition to reality is made by theoretical physics, which correlates symbols to observed phenomena. Where this can be done hidden structures are coordinated to phenomena; these very structures are regarded by the physicist as the objective reality lying behind the subjective phenomena. ... I am not afraid of identifying such well-defined structures with Kant's "thing in itself". The objection ... [that] the existence of the "thing in itself" is postulated because one needs an external cause to understand why different individuals experience "the same" phenomena; but the category of causality has meaning only within the domain of phenomena ... [has] no validity from our point of view ... the concept of causality is

> a residue of former ways of thinking and is replaced by the process of coordination as described before. This procedure leads to structures which are communicable, controllable, hence objective. It is justified to call these by the old term "thing in itself". They are pure form, void of all sensual qualities. That is all we can wish or expect. ... one can agree with Hegel that they are a "perfect abstractum". But that they are perfectly empty does not fit the facts. Remember what practical use can be made of them. ...[21]

As it happened, some years before writing this statement, Born had written to Bohr that he "disliked thoroughly" the instrumentalist standpoint of those physicists and philosophers who hold that the goal of science is a "purely observational description in which one does not inquire what there is behind the phenomenon".[22] Bohr replied to Born:

> Indeed, it is difficult for me to associate any meaning with the question of what is behind the phenomena, beyond the correspondence features of the formalism which itself represents a mathematical generalization of the classical physical theories permitting, within its scope, predictions of all well defined observations which can be obtained by any conceivable experimental arrangement.[23]

In his reply to Bohr's letter Born explained:

> What I meant by "behind the phenomena" is in mathematical language just "invariants" in the most general sense of the word. The various aspects of phenomena which we consider in quantum mechanics have also a theory of "invariants", or in less learned language, common features which do not depend on the aspect, and it is this which I would like to preserve as something beyond our direct experience. ... If one does not accept such a standpoint, it appears to me that one accepts a hypersubjective or solipsistic standpoint, and that one resigns oneself to answering any questions about why one is investigating the world at all.[24]

To this explanation on Born's part, Bohr replied:

> I owe you, of course, an apology for my remarks as to the question of what is behind the phenomena. ... As you express your views in the letter, I agree entirely and had myself the same attitude when in my letter I spoke of a consistent abstract generalization of classical mechanics and electrodynamics interpretable only on correspondence lines.[25]

Thus here we see concrete evidence of the fact that while Bohr himself did not take complementarity as a springboard to conclusions about the nature of a reality behind the phenomena, he did not hold the phenomenalist view that this notion refers to nothing that enters into scientific description. Indeed, he states quite simply that he "agrees entirely" with Born's position that the mathematical symbols of the formalism which allow us to predict the phenomenal appearances of an atomic object express what we know about this reality behind the phenomena.

An extended analogy involving objects more familiar than those of quantum physics will help give some concrete expression to this rather abstract analysis of our knowledge of an independent reality as seen from the framework of complementarity. Imagine the object of our intended description is an individual personality and that our goal is the *objective* description of this personality through the behavior the individual manifests in different contexts. Traditionally this would have been done by treating the object of description as a sort of entity, a "self" or "ego", to which various personality traits would belong as properties possessed by this entity over periods of time, though changeable over a long run through certain causal influences. The existence of such a metaphysical "self" is of course what a purely empirical epistemology cannot support, but its abandonment seems to rob the person of an identity which each individual would claim on the basis of his or her own experiences. The viewpoint of complementarity suggests that this sort of treatment has gotten into difficulties because it has taken a problem in the use of descriptive concepts and turned it into a metaphysical issue on the existence or non-existence of some alleged entity with certain properties to which the attempted description refers. Instead, complementarity presents an alternative way to view this descriptive task.

Consider a concrete individual, Smith, whose behavior is limited to phenomenal appearances arising in his interactions with three other "systems", say his wife, his children, and his supervisor. Furthermore, suppose that in each individual interaction it is possible to describe Smith's phenomenal appearances with total consistency, but that the three different descriptions of these phenomena require terms which when predicated of one and the same object would be, by definition, contradictory. Assume that these descriptions report that Smith is loving and competent in interactions with his wife, imperious and arbitrary with his children, and meek and indecisive with his supervisor. Certainly if one naively thought that these descriptive terms are intended to refer to properties possessed by a single "object", Smith's "ego", the resulting description of such an "object" would appear to attribute contradictory properties to Smith. Thus we have learned implicitly to treat such behavioral characteristics as powers or tendencies to behave in specific ways when interacting with certain sorts of other entities. The problem is, does such a "description" give us anything more than a set of different subjective responses to the way (the phenomenal) Smith behaves? What, we are tempted to ask, are the properties of the "real" Smith, as he would be independently of his interactions with these other objects, such that he behaves in these ways when interacting with these objects?

Bohr would approach this problem by pointing out that our task is to formulate a conceptual framework and a theoretical formalism which will render possible unambiguous communication about the phenomena through which we say we observe Smith. For this task, we have as our given data only the experienced phenomena as communicated in the *subjective* reports of the three "observing systems". Thus for example the supervisor describes a (phenomenal) person possessing the property of indecisiveness. If this property which truly describes the phenomenal appearance of Smith as an object of description is taken to refer to a property possessed by the "real" Smith, by Smith's ego, the resulting description cannot be reconciled with Smith's wife's description of her husband's ego as possessing the property of competence. These communications are "ambiguous" because they are describing different *phenomenal* objects. But by ignoring the full circumstances of the observational interactions, one might assume they both refer to "the same object".

In order to remove this ambiguity, we must develop a descriptive formalism which will allow us to describe Smith's ego in a way which will predict not only that Smith will behave indecisively before his supervisor and competently before his wife, but also, prior to the observation of it, that he will behave imperiously before his children. We need not change the terms of our description of the *phenomenal* Smith, for indeed they describe the phenomenal object unambiguously. But we must understand that such descriptive concepts which are given empirical reference to the phenomenal object are unambiguous only when describing specific interactions between Smith and other persons. These concepts, therefore, are not defined for reference to the independently existing Smith, isolated from his interactions with others. In trying to isolate Smith's behavior from the phenomena in which that behavior is manifested, we create an abstraction for which there can be no empirical evidence. Smith and his interactions with other persons have an "individuality" which makes ambiguous any attempt to describe a phenomenal Smith, without specifying the full circumstances of the observational interaction in which this phenomenal Smith was observed.

Once this epistemological lesson is learned, it is possible to see how Smith may be described "objectively" in Bohr's sense. Let us say that we start by describing only that appearance which Smith presents to his supervisor. Clearly to mistake such a description of Smith as a description of Smith's "ego" (construed as the independent reality lying behind Smith's phenomenal appearances) is to mistake a phenomenal description for a description of an independent reality. If we add the knowledge of how he behaves before his wife and children, then we are aware of several different phenomenal Smiths. At this point we would have arrived at the state of atomic physics in 1924, prior to the appearance of Heisenberg's and Schrödinger's formalisms. The next step, then, would be to develop a theory of personality or behavior according to which Smith may be characterized by purely "abstract symbols" which are defined theoretically but have no direct empirical reference. According to such a characterization of Smith, we can predict his behavior successfully in these and other observing interactions. Clearly in this last step our knowledge of Smith has advanced a great deal. We now have a theoretical formalism for co-ordinating different phenomenal descriptions of Smith's behavioral appearances as "complementary" descriptions of the "same" person or self. But what have we learned about Smith? More different appearances of Smith-behavior? No, though we may discover more phenomenal Smiths, they are significant only by way of corroborating and instilling greater confidence in our theory. Thus it must be that, if we have learned more about Smith, it cannot be the *phenomenal* Smith, for this new knowledge is expressed by terms in the formalism which have no empirical reference. So it must be of some other object, and the knowledge about this object, contained in the theory, is what we mean by having knowledge of the grounds of the different phenomenal appearances of Smith-behavior. This is the reason that justifies our claim that, by being able to relate various different descriptions of different phenomena as appearances of the same object, we have learned something about that object which lies behind these behavioral appearances.

4. The Ontological Status of an Independent Physical Reality

Inasmuch as complementarity aspires to no more (nor less) than presenting a conceptual framework for consistently and objectively describing phenomena as observational interactions which have an individuality that prohibits subdividing them into separate interacting systems, it is not to be expected that it specify the ontological status of the object of its descriptions any more than does the traditional framework of classical physics. In other words, complementarity is not an ontology. However, it does not follow that it has no consequences for ontology, for indeed any conceptual framework appropriate for describing nature has historically been an important datum for a theory of reality. Thus, following Bohr, in this section I will put forward only a minimum of positive ontological claims. However, I will also concentrate on pointing out, as Bohr did not, the extent to which complementarity is incompatible with the traditional assumption that reality can be described in terms of the properties possessed by substantial entities.

Of course complementarity cannot claim to demonstrate that reality is of such and such a nature any more than classical mechanism could demonstrate that reality is composed of mass-particles and energy fields. The ontological hypothesis that reality is so constituted has its credibility because of its conformity with the framework of classical physics and because of the success of classical physics in describing a wide variety of phenomena. If that framework must be replaced, as Bohr has argued, Then this hypothesis becomes untenable, and it is necessary to develop an alternative ontology. In any case, we must recognize that the ontological presuppositions we adopt are not *demonstrated* by scientific theory, but are hypothesized in order to justify that a theory in a given framework permits an acceptable and complete description of the nature of physical reality. Nevertheless, our allegiance to the ontology of classical physics tends to be so strong that, failing to be able to demonstrate that the *classical realist* ontology of physical systems (existing as independently real entities with the properties which define their classical mechanical states) is compatible with the quantum postulate, we mistakenly conclude that when we accept quantum mechanics, we resign ourselves from *any* description of an independent reality and fall into an anti-realist view.

The ontological significance of complementarity requires that a new account of the natural world alters the notion of the "object" of description in physics, the "physical system". In the classical mechanistic view there are 1) the theoretical representation of the classical mechanical state of the system defined in terms of the parameters of space, time, momentum, and energy, 2) the phenomenal object or system as observed characterized in terms of those observable properties which give direct empirical reference to some of the terms of the theoretical representation, and 3) an independent physical reality, characterized as an entity possessing those properties which are held to cause the observable properties of the phenomenal object and which in turn correspond to the terms that define the classical mechanical state in the theoretical representation of that state. The first allows us to predict the characteristics of the second, which in turn is regarded as the phenomenal effect of the third. It is this reason which justifies describing the object as an independent reality in terms of the properties of the phenomenal object. This way of describing nature allowed one to "picture" the state of a physical system as it exists isolated from any observation of it.

In contrast, complementarity divides the epistemic content of the theoretical understanding of nature into two types of knowledge: a) a description of a phenomenon as an observational interaction in which the whole interaction is partitioned into observing system and observed phenomenal object, and b) the purely formal theoretical structures which symbolize the quantum mechanical state of an independent object lying behind the phenomenon in terms that allow for co-ordinating different phenomenal appearances as appearances of the same object. Because the theoretical structures which symbolize this state are purely "abstract", having no reference to anything directly observable in experience, the independent physical reality which lies behind the phenomenal objects cannot be described as possessing properties corresponding to the properties used for describing phenomenal objects. Thus the use of the formalism to picture the state of the system isolated from observations refers an abstraction, not a model of an independent physical reality.

But upholding a realistic interpretation of quantum theory does not require that we regard the atomic systems which are the objects of our knowledge in quantum theory as composed of substances possessing properties like those of the phenomenal objects which give empirical reference to the theoretical descriptive terms. A realistic interpretation of classical physics could accord the object of scientific knowledge the ontological status of a "material substance" which was pictured as possessing the same primary properties as observed in phenomenal objects. Bohr has shown that such a view becomes impossible in complementarity. However, Bohr equally rejects the idealistic notion that the object of scientific knowledge has the ontological status of an object of consciousness. Thus he points out that complementarity ...

> ... offers a logical means of comprehending wider fields of experience, necessitating proper attention to the placing of the subject–object separation. Since, in philosophical literature, reference is sometimes made to different levels of objectivity or subjectivity or even reality, it may

> be stressed that the notion of an ultimate subject as well as conceptions like realism and idealism find no place in objective description as we have defined it; but this circumstance of course does not imply any limitation of the scope of the enquiry with which we are concerned.[26]

Bohr's point is that traditional conceptions of what he calls an "ultimate subject" are based on the assumption that such a subject is to be described by the same approach as that which is unambiguous for describing phenomenal objects, for the phenomenal object has a definite set of characteristics only in a description in which the subject–object distinction may be shifted. Thus, like the Licentiate in Møller's tale, the traditional ontology has sought to make an object out of the cognizing subject only to discover that in doing so, we have shifted the subject–object distinction without noticing it. Bohr might just as well have said the same about an "ultimate object", which is actually more in keeping with his concern to point out his differences with realism as classically understood. Thus in the example of seeing a red apple, the "object" of description may be a fruit, the light radiation entering my eye, the simulation of the optic nerve, or the presentation in my consciousness of something variously termed an "idea", "perception", "impression", or "representation". The philosophical assumption that somehow *one* of these "objects" is the "ultimate object" in the sense of *the* cause of the experience, and all the others are its effects, is a mistake bound up with the representationalistic epistemology which underlies classical realism. According to this view the description of observation describes the causal *result* of a physical interaction on a subjective consciousness, rather than the interaction itself. This mistake fuels the ontological efforts of idealism and classical realism.

The fact that idealism and classical realism are not viable options for the description of nature in complementarity may suggest a phenomenalist reading of Bohr's viewpoint. However, the choice between idealism and classical realism arises only if we assume that in order to communicate knowledge of a real world behind the phenomena, a theory must establish a representationalistic relationship between its terms and the properties of the object of knowledge. The fact that complementarity rejects this assumption means that we need not answer the question of which of these alternatives to choose. Furthermore, the restriction of ontology to a substance–property view of reality results from this same assumption, for the substance–property relation is inherent in the description of phenomenal objects and thus by the assumption of representationalism must be recapitulated onto the level of the reality which lies behind these phenomena.

For complementarity the question between idealism and classical realism is pointless. For example, to argue, as does the idealist, that the ontological status of an independent reality is the same as that of "ideas" would be a complete *non sequitur* based on premises regarding the description of *phenomenal* objects. According to complementarity, that type of description is *not* to be applied to an independent physical reality. For any empiricistic epistemology, to have an "idea" with any content is to have an idea of something that could be experienced; thus ideas must be regarded as referring to phenomenal objects. To hold that reality has the ontological status of ideas is, therefore, to hold that reality can be described the same way as phenomenal objects. But this is exactly what complementarity denies. On the basis of the quantum theoretical description of nature viewed from complementarity, it is surely misleading to say that we know reality only in the way we form a conscious representation of *phenomenal* objects, hence reality must *be* of the same status as that which is present in consciousness, namely ideas or perceptions.

The classical realist argues that the ontological status of independently real objects is the same as that of "objects" represented by the classical concept of the state of the isolated system, namely that which is defined as possessing the classical mechanical properties. Bohr's point is not only that this position simply invests an abstraction with concrete status, but also that in light of the discovery of wave–particle dualism, we have definite grounds to reject any attempt to make this identification. Once we abandon this view, the apparently scientific foundation for classical realism falls.

Since the theoretical structures of quantum theory provide no justification for concluding that an independent physical reality possesses the spatio-temporal properties of a classical entity, the "picture" of an isolated system in space and time cannot be used to characterize the ontological status of an independent reality within the framework of complementarity. Thus the quantization of interaction leads contemporary physics away from the classical realist conception of the atomic domain. Instead, it indicates that we can describe such a reality in terms of its power to produce the various different observational interactions described by the theory as providing complementary evidence about the same object. This independent reality which Bohr *presupposes* in all of his talk about observation as interaction must be described in terms of its power to structure phenomenal interactions in such a fashion that what is observable is predictable on the basis of theoretical knowledge. Of course, it is essential to empirical knowledge that such predictions can be described in terms which refer to properties of phenomena found in experience, specifically those which provide the empirical reference for the classical descriptive concepts. Thus an independent reality must be characterized as having the power to produce phenomena in interactions such that these phenomena are described by the phenomenal observables to which the classical concepts refer. The description of phenomena as an indivisible interaction requires that, in characterizing the reality behind the phenomena through its activity in originating phenomena, we attribute to this activity an individuality which means that the distinction between observing system and phenomenal object is an arbitrary one made for the purpose of securing an unambiguous description. Since this distinction could have been drawn otherwise, the individuality of the interaction implies that we cannot regard any *phenomenal* object so described as corresponding in its properties to the properties of an independently real object.

We may convey the force of these conclusions by saying either that the traditional substance–property ontology is abandoned, or that it is radically restructured. If one assumes that in a substance–property ontology all ontological claims assert that a particular substance possesses certain determinate properties, where properties are regarded as directly causing phenomenal observables, then complementarity is inconsistent with a substance–property ontology so defined. However, if one defines "substance" to mean any subject of predication, and defines "properties" to include the powers to produce certain phenomena in specific circumstances, then complementarity can be considered as consistent with a substance–property ontology in this sense. The important point is not the terminology but the recognition that what complementarity allows us to say about the reality which lies behind the phenomena is that it has the characteristics of being able to produce different sorts of phenomena in different sorts of interactions, and the way these phenomena are described cannot be used to characterize the reality which causes them. In one sense it would seem that such a characterization would be more adequately expressed by speaking of the powers of an activity instead of the properties of a substance. However, in another sense, properties have always been regarded as powers and substances as things acting upon observers; in this latter sense complementarity is a continuation of substance–property ontology.

Freeing the concept of a reality behind the phenomena from the presupposition that such a reality must be described as a substance possessing properties removes the temptation to concede that a realistic defense of quantum theory is impossible. But our language is so influenced by the character of the descriptions of phenomenal objects that a verbal communication referring to the reality symbolized in the formal mathematical structures of the theory is quite difficult. Nevertheless, whatever else the nature of an independent reality might entail, the assumption that such a reality is endowed with the power to interact with other physical systems to produce the phenomena we observe is directly required by complementarity. The formal theoretical structures from which we can predict descriptions of observed phenomena provide a symbolic expression for the characteristics of such an activity.

We should not misunderstand this talk of the theoretical formalism as providing a symbolic understanding of the cause of phenomena to imply that the formal abstract parameters of the theory correspond to determinate properties of the independently real objects. They are only properties of theoretical structures. These structures are symbols allowing us to make statements about the grounds of atomic phenomena, not because of some representational correspondence between parameters and properties, but because, using such symbols, we are able to relate various phenomenal descriptions as complementary phenomenal appearances of the same object.

A set of complementary descriptions of different phenomena (theoretically structured so that they are regarded as exhausting all that is empirically observable about the same object) provides the justification for regarding this object as the ontological grounds of the appearances of these phenomena. These theoretical structures stipulate the notion of identity implicit in these claims. Nevertheless, it is worthwhile to note that the self-identity stipulated by the theory is not given ontological significance through appeal to a "material substance" (as classically conceived) persisting through a variety of forms. Any ontology consistent with complementarity must recognize that an ontology of the grounds of phenomena can be specified only by defining it as symbolized by the theoretical structures which allow us to predict what will be experienced as phenomena.

Thus complementarity suggests developing an ontological conception of the independent reality which lies behind the phenomena based on the notion of an independent but interacting physical reality, the grounds of experience, but not describable by the terms of experience. Beyond such a general characterization, little else can be said regarding the positive nature of such an ontology. Yet these few indications should suggest the powerful ontological themes which can be developed once the road is clear to understanding that within the framework of complementarity science can retain a respectable measure of its traditional claim to providing knowledge of objects as they are independent of our observations of them.

5. Final Comments

Bohr loved to make his points through humorous anecdotes; one such is reported by Aage Petersen, a student and close associate of Bohr during his later years. The citizens of a small Jewish community sent one of their brightest young men to hear the teachings of a famous rabbi in a nearby city. The young man returned with the following report:

> The rabbi spoke three times. The first talk was brilliant; clear and simple. I understood every word. The second was even better; deep and subtle. I didn't understand much, but the rabbi understood all of it. The third was by far the finest; a great and unforgettable experience. I understood nothing, and the rabbi didn't understand much either.[27]

We may analogize Bohr to the rabbi and the young man to his audiences. With respect to interpreting quantum physics, both Bohr and his sympathetically inclined listeners seem to have understood the main teachings of complementarity. With respect to his general epistemological lesson, it seems fair to say that Bohr had a clear understanding of his message, but he lost a great part of his audience. However, with respect to the ontological revisions in the concept of physical reality which are implicitly required by the framework of complementarity, we must conclude that neither Bohr nor his audiences understood the issues clearly, though there was a general feeling that profound ideas were involved.

By examining complementarity in the context of its historical and conceptual genesis as it took place in Bohr's own thought, we have seen how from the very start it was intended to be a framework which would replace that of classical mechanism. Although Bohr designed this framework to remove the paradoxes of quantum theory, it was not intended as just an interpretation of quantum mechanics. In generalizing the older framework to produce a new one compatible with the quantum postulate, Bohr has shown that the older framework presupposed that interactions take place continuously, a presupposition that was discarded at the beginning of the quantum revolution. Thus, a "more general" case would not make this special presupposition. The result is that we must relinquish the classical view that the two ideals of observation and definition could both be attained in a single description. Instead, we must recognize, that the goal of observation, a space–time description of the physical system, is complementary to the goal of defining the state of a system through determining its causal effect on the observing system by means of the conservation principles. The physical conditions required for phenomena which theory describes as an observational interaction determining spatio-temporal parameters exclude the conditions necessary for phenomena which are described as interactions determining momentum or energy through the conservation principles. Thus the phenomena described in these different observational interactions are complementary phenomenal appearances which together provide all that can be known about the independently real same object which lies behind these phenomenal appearances. The descriptions produced by applying these complementary modes of description will not cause inconsistencies as long as we recognize that what is described in each case is the whole phenomenon treated as an observational interaction in which we make a clear but arbitrary distinction between observing system and observed phenomenal object. In order to make good its claim to objectivity, complementarity must show how these restrictions on the use of classical concepts make possible unambiguous descriptions of the relevant phenomena. At the same time, we have also shown that in basing its position on the empirical discovery expressed by the quantum postulate, complementarity must not discard the concept of an independent physical reality in the atomic domain, which in interacting with observing systems produces the phenomena that confirm quantum theory. Therefore, Bohr must avert the conclusion that his epistemological lesson leads to an anti-realistic interpretation of quantum theory. For this reason, it is necessary to add to Bohr's view the ontological position that by combining different phenomenal descriptions through the formal structures of quantum theory, as different phenomenal appearances of the same object, it is possible to have theoretical knowledge of the ontologically independent objects that are regarded as the reality lying behind the phenomena. The ontology implied by this interpretation of Bohr's message characterizes physical objects through their powers to appear in different phenomenal manifestations rather than through determinate properties corresponding to those of phenomenal objects as was held in the classical framework. The case is then made that, within the framework of complementarity, it is possible to preserve a realistic understanding of science and accept the completeness of quantum theory only by revising our understanding of the nature of an independent physical reality and how we can have knowledge of it.

Bohr's published works are such that it is easy to make out the worst possible case for complementarity. Basically Bohr's minimal acquaintance with the vocabulary and positions of traditional philosophy led him to ill-advised expressions. Furthermore, his reluctance to express an ontological view to buttress his claims about observation as interaction leaves complementarity open to criticisms often leveled on the misinterpretation that it is a defense of an anti-realistic interpretation of quantum theory and science in general.

In this analysis, however, I have determined to make the best possible case for the important and philosophically ignored insights contained within the complementarity viewpoint. By doing this, I have shown how this framework can be regarded as one which is consistent, coherent, and comprehensive enough to include all of natural science. Furthermore, I hope that it appears that this framework represents an improvement on that of mechanism, inasmuch as, if we accept contemporary physics, we must reject the presuppositions on which the older framework rests. But we must not forget that all of the successful descriptions in the classical framework are retained as special case descriptions within complementarity.

However, to claim that complementarity is the final framework for all empirical knowledge would reveal a failure to appreciate Bohr's epistemological lesson. Indeed, Bohr clearly held that "we must be prepared for a more comprehensive generalization of the complementary mode of description which will demand a still more radical renunciation of the usual claims of so-called visualization".[28] Thus with Bohr we should be prepared to expect that the further exploration of nature will reveal new phenomena that bring to light the need to revise other presuppositions on which our current viewpoint rests, thereby entailing the need for a more general framework in which complementarity may well appear as a special case. Thus we should be prepared for surprises, as Bohr was fond of saying, and rest assured that future frameworks will once again alter our understanding of physical reality and how we know it. In a certain sense, Bohr's deepest lesson, the one he most earnestly strived to convey in all of his physics and philosophy, is the need to inculcate this state of mind as the necessary condition for progress in scientific understanding. In that sense, in Bohr's own words, his philosophical contribution "may be summarized as the endeavor to achieve a harmonious comprehension of ever widening aspects of our situation, recognizing that no experience is definable without a logical frame and that any apparent disharmony can be removed only by an appropriate widening of the conceptual framework".[29]

NOTES

CHAPTER ONE

1 *Archive for the History of Quantum Physics*, Interview with Professor Niels Bohr, conducted by Thomas Kuhn *et al.* (November 17, 1962), transcript, p. 3. Hereafter this source will be cited as "Last Interview", and reference to the Archives will be cited as "*AHQP*". An inventory of the holdings of *AHQP* as of 1967 may be found in T.S. Kuhn *et al.*, *Sources for History of Quantum Physics* (Philadelphia: American Philosophical Society, 1967).
2 Niels Bohr, "Can the Quantum-Mechanical Description of Physical Reality be Considered Complete?", *Physical Review 48* (1935), 696.
3 *Ibid.*, p.697.
4 *Niels Bohr Archives*, Karl Taylor Compton Lectures, delivered at Massachusetts Institute of Technology, November 21, 1957; Fifth Lecture, unpublished transcript, p. 6. Hereafter reference to the Niels Bohr Archives will be cited as "*NBA*"; a partial listing of the contents of the archive, which is held in conjunction with *AHQP*, is found in Kuhn *et al.*, *Sources*, 1967.
5 *NBA*, Gifford Lectures, delivered in Edinburgh, October, 1949; Eighth Lecture, unpublished transcript, p. 2.
6 Niels Bohr, "Causality and Complementarity", *Philosophy of Science 4* (1937), 289.
7 *Ibid.*, p. 290.
8 *Idem.*
9 Niels Bohr, "Analysis and Synthesis in Science", Introduction to *Encyclopedia and Unified Science, Foundations of the Unity of Science, Vol. I. No. 1, International Encyclopedia of Unified Science*, ed. by Otto Neurath (Chicago: University of Chicago Press, 1938), p. 28. *Cf.* also the essay "Unity of Knowledge", in Niels Bohr, *Atomic Physics and Human Knowledge* (New York: John Wiley and Sons, Inc., 1958), pp. 67–68.
10 Niels Bohr, "Causality and Complementarity", *Phil. Sci. 4*, 289–290.
11 Niels Bohr, "Light and Life", in *Atomic Physics and Human Knowledge*, p. 3.
12 Niels Bohr, "Unity of Knowledge", in *Atomic Physcis and Human Knowledge*, p. 68.
13 Niels Bohr, "Physical Science and Man's Position", *Philosophy Today* (1957), 67.
14 Niels Bohr, "Newton's Principles and Modern Atomic Mechanics", in The Royal Society, *Newton Tercentenary Celebrations* (Cambridge: The University Press, 1947), p. 57.
15 Niels Bohr, "Atomic Physics and International Cooperation", *Proceedings of the American Philosophical Society 91* (April, 1947), 138.
16 Niels Bohr, Physical Science and the Study of Religions, in *Studia Orientalia Ioanni Petersen* (Copenhagen: Einar Munksgaard, 1953), p. 386.
17 Niels Bohr, "Unity of Knowledge", in *Atomic Physics and Human Knowledge*, pp. 67–68.
18 *NBA*, "Atoms and Human Knowledge—Nicola Tesla", reprint of a lecture given at the Nicola Tesla Conference, 1956.
19 I do not intend this distinction to suggest that certain descriptive terms have a meaning apart from the framework in which they are employed. Indeed Bohr would hold that an observational/theoretical distinction cannot have meaning apart from the framework within which it is employed. My only point is that all descriptive terms cannot be defined solely in terms of each other (implicitly); in an empirical science some terms must be made to refer to some

element of experience, even though such a reference is not possible apart from the framework of presuppositions governing the correct use of such concepts.

20 Heisenberg's principle goes under both names, "uncertainty" and "indeterminacy", in English. Though these terms are by no means synonymous, in choosing "uncertainty principle" throughout the remainder of this work I do not mean to advocate any particular interpretation of this theoretical principle. My choice here is dictated by popular usage (which is also Heisenberg's and Bohr's on most occasions when speaking English) which seems to have settled upon "uncertainty". On the difference between these terms, *cf.*, Max Jammer, *The Philosophy of Quantum Mechanics* (New York: John Wiley & Sons, 1974), pp. 61–62.

21 I am aware, of course, that this view is controversial; however, it is necessary to postpone my defense of Bohr's realism until after we have analyzed complementarity itself. *Cf.*, Chapters Seven and Eight below.

22 Bub holds that Bohr was "a remarkably successful propagandist" for "an approach to the problem of knowledge that had fascinated him since youth" and that he created complementarity to fit this preconceived philosphical outlook. Arguing from second-hand attributions of quotes to Bohr, without direct reference to Bohr's own writings, he concludes that the views expressed "reveal Bohr's philosophical hang-ups, no more. The careful phraseology of complementarity, drawing on this resevoir, endows an unacceptable theory of measurement with mystery and apparent profundity, where clarity would reveal an unsolved problem." Jeffrey Bub, *The Interpretation of Quantum Mechanics* (Dordrecht and Boston: D. Reidel, 1974), pp. 45–46. Taking a different tack, Lewis Feuer also attempts to portray Bohr as committed to philosophical preconceptions which give complementarity its character. *Cf.*, Lewis Feuer, *Einstein and the Generations of Science* (New York: Basic Books, 1974), pp. 121–126. In both cases arguments are based on the slimmest possible historical documentation; certainly a thinker of Bohr's caliber deserves better than this.

23 *Cf.*, P.K. Feyerabend, "Niels Bohr's Interpretation of the Quantum Theory", in *Current Issues in the Philosophy of Science*, ed. by H. Feigl and G. Maxwell (New York: Holt, Rinehart and Winston, 1961); and "Problems of Microphysics", in *Frontiers of Science and Philosophy*, ed. by R.G. Colodny (Pittsburgh: University of Pittsburgh Press, 1962).

24 Bohr's association with positivism seems credible in the light of his contribution of a short introductory statement to Vol. I, No.1 of the *Encyclopedia of Unified Science*, a project motivated by positivist philosophers of science. But unless we apply the principle of guilt by association, it is clear from all of his writings that Bohr never adopted a positivist model of science or a verificationist criterion of meaning. Patrick Heelan seems to simply assume that Bohr's view is a form of "facile positivism", *cf.*, *Quantum Mechanics and Objectivity* (The Hague: Martinus Nijhoff, 1965), p. ix. But the most vocal sponsers of the positivist and subjectivist reading of Bohr are Karl Popper and Mario Bunge; *cf.* for example, Karl Popper, "Quantum Mechanics without the Observer", and Mario Bunge, "The Turn of the Tide", both in *Quantum Theory and Reality*, ed. by Mario Bunge (New York: Springer Verlag, 1967). Michael Audi nicely explodes these misinterpretations in *The Interpretation of Quantum Mechanics* (Chicago: The University of Chicago Press, 1973), pp. 15–27.

25 Bohr's record of his debate with Einstein appears in "Discussions with Einstein on Epistemological Problems in Atomic Physics", in *Albert Einstein: Philosopher–Scientist, Vol. I*, ed. by P.A. Schilpp (New York: Harper & Row, 1949), pp. 199–242. An account which does a superb job of showing that the debate involved radically opposing conceptions of reality is C.A. Hooker, "The Nature of Quantum Mechanical Reality: Einstein Versus Bohr", in *Para-*

digms and Paradoxes, ed. by R.G. Colodny (Pittsburgh: University of Pittsburgh Press, 1972), pp. 67–302.

26 Jammer, *The Philosophy of Quantum Mechanics*, p. 247.

27 *Ibid.*, pp. 247–248.

28 Bohr would agree with the view that the descriptive concepts are "theory-laden" if "theory" is taken to include the conceptual framework within which a mathematical formalism is interpreted as providing an account of experienced phenomena. Nevertheless, his position is that the exploration of nature uncovers new phenomena which affect which conceptual framework will permit an unambiguous description of such phenomena. *Cf.*, N.R. Hanson, *Patterns of Discovery* (Cambridge: Cambridge University Press, 1958); T.S. Kuhn, *The Structure of Scientific Revolutions, Second Edition* (Chicago: The University of Chicago Press, 1970); and P.K. Feyerabend, "Explanation, Reduction, and Empiricism", in *Minnesota Studies in Philosophy of Science, Vol. 3*, ed. by H. Feigl and G. Maxwell (Minneapolis: University of Minnesota Press, 1962), pp. 28–97.

CHAPTER TWO

1 *AHQP*, "Last Interview", p. 2.

2 David Jens Adler, "Childhood and Youth", in *Niels Bohr: His Life and Work as Seen by His Friends and Colleagues* (New York: John Wiley and Sons, 1964), p. 25.

3 *Ibid.*, p. 15.

4 Niels Bohr, "On the Constitution of Atoms and Molecules", *Philosophical Magazine 26*, 1–25, 476–502, 857–875 (1913). These articles have been reprinted with an illuminating introduction by Leon Rosenfeld in Niels Bohr, *On the Constitution of Atoms and Molecules* (Copenhagen: Munksgaard, 1963).

5 Oskar Klein, "Glimpses of Niels Bohr as Scientist and Thinker", in Rozental (ed.), *Niels Bohr*, pp. 84–85.

6 *AHQP*, Interview with Werner Heisenberg, conducted by T.S. Kuhn (Munich, 19 Feb. 1963), Tape 50b, transcript pp. 7–8.

7 Niels Bohr, *Atomic Theory and the Description of Nature* (Cambridge: The University Press, 1934, reprinted 1961).

8 Niels Bohr, *The Times*, August 11, 1945. *Cf.* also, Niels Bohr, "A Challenge to Civilization", *Science 102* (1945), 363.

9 Niels Bohr, "An Open Letter to the United Nations", reprinted in Rozental (ed.) *Niels Bohr*, pp. 340–352. In this document Bohr details his relation to the atomic bomb project and his early attempts to caution political leaders of its implications.

10 Niels Bohr, *Atomic Physics and Human Knowledge* (New York, John Wiley and Sons, 1958) and *Essays 1958–1962 on Atomic Physics and Human Knowledge* (New York: John Wiley and Sons, 1963).

11 Klein, "Glimpses", in Rozental (ed.) *Niels Bohr*, p. 79.

12 Leon Rosenfeld, *Niels Bohr: An Essay* (Amsterdam: North Holland, 1961), p. 5.

13 *Idem.*

14 *Ibid.*, pp. 5–6.

15 Leon Rosenfeld and Erik Rüdinger, "The Decisive Years: 1911–1918", in Rozental (ed.), *Niels Bohr*, p. 54.

16 Werner Heisenberg, *Physics and Beyond* (New York: Harper & Row, 1961), p. 38; and "Quantum Theory and Its Interpretation", in Rozental (ed.), *Niels Bohr*, pp. 94–95, *passim.*

17 Rosenfeld, *Niels Bohr: An Essay*, p. 8

18 Leon Rosenfeld, "Niels Bohr in the Thirties", in Rozental (ed.), *Niels Bohr*, p. 117.

19 Rosenfeld, *Niels Bohr: An Essay*, pp. 8–9.

20 Max Jammer, *The Conceptual Development of Quantum Mechanics* (New York: McGraw-Hill, 1966), p. 166.

21 For a prudent admonition in this respect, *cf.*, David Favrholdt, "Niels Bohr and Danish Philosophy", *Danish Yearbook in Philosophy 13* (1976), 206–222.

22 *Ibid.*, p. 206.

23 *AHQP*, "Last Interview", p. 3.

24 *AHQP*, Interview with Oskar Klein, conducted by T.S. Kuhn and J.L. Heilbron (Copenhagen, 16 Jul. 1963), Tape 83b, transcript p. 9.

25 Gerald Holton, "The Roots of Complementarity", *Daedalus 99* (Fall 1970), 1035; reprinted in G. Holton, *Thematic Origins of Scientific Thought* (Cambridge, Mass.: Harvard University Press, 1973), p. 138. *Cf.* also, Favrholdt, "Niels Bohr and Danish Philosophy", *Danish Yearbook 13*, 217.

26 Niels Bohr, "Physical Science and the Problem of Life", in *Atomic Physics and Human Knowledge*, p. 96.

27 David Jens Adler, "Childhood and Youth", in Rozental (ed.) *Niels Bohr*, pp. 27–28.

28 Holton, "The Roots of Complementarity", *Daedalus 99*, 1045; *Thematic Origins*, p. 146. An extended attempt to make a case for the Kierkegaardian influence appears in Lewis Feuer, *Einstein and the Generations of Science*, pp. 121–126. In my judgment such attempts to read ideas into Bohr's unconscious motivations are not only impossible to authenticate but essentially rest on gross misunderstandings of complementarity that have done more harm than good in advancing the understanding of Bohr's philosophy; Feuer's speculative fantasy here brings no credit to the genre of "psychobiography". A far more balanced position is taken by ones who knew him well, in this respect, *cf.*, J. Rud Nielsen, "Memories of Niels Bohr", *Physics Today 16* (1963), 22–30.

29 *NBA*, "Høffdings 85 ars dag", *Berlingske Tidende*, March 10, 1928 [Copenhagen daily newspaper] and "Harald Høffding—Memorial Tribute", 1932, privately published. *Cf.* Favrholdt, "Niels Bohr and Danish Philosophy", *Danish Yearbook 13*, 214 for a partial translation of the former.

30 *AHQP*, "Last Interview", p. 5.

31 *Cf.* Holton, *Daedalus 99*, 1035; *Thematic Origins*, p. 138; Favrholdt, *Danish Yearbook 13*, 216–217; Jammer, *Conceptual Development*, pp. 177–179; Henry Pierce Stapp, "The Copenhagen Interpretation", *American Journal of Physics 40* (August 1972), 1104–1105, 1113–1115; and K.M. Meyer-Abich, *Korrespondenz, Individualität, und Komplementarität* (Wiesbaden: Franz Steiner Verlag, 1965), pp. 133–136.

32 Holton, *Daedalus 99*, 1035; *Thematic Origins*, p. 139.

33 Letter from Niels Bohr to Harald Bohr, 26 June 1910, quoted in *Collected Works of Niels Bohr, Vol. 1*, ed. by J. Rud Nielsen (Amsterdam: North Holland, 1972), p. 513.

34 Feuer, *Generations of Science*, p. 128.

35 Niels Bohr, "The Unity of Human Knowledge—1960", *Essays 1958–1962*, p. 13.

36 Rosenfeld, "Niels Bohr in the Thirties", in Rozental (ed.), *Niels Bohr*, p. 121.

37 Leon Rosenfeld, "Niels Bohr's Contributions to Epistemology", *Physics Today 16* (1963), 48.

38 Rosenfeld, "Niels Bohr in the Thirties", in Rozental (ed.) *Niels Bohr*, p. 121.

39 Bohr, "The Unity of Human Knowledge—1960", *Essays 1958–1962*, p. 13.
40 *Idem.*
41 Hans Bohr, "My Father", in Rozental (ed.), *Niels Bohr*, pp. 327–328.
42 Quoted by Holton, *Daedalus 99*, 1043; *Thematic Origins*, p. 148.
43 Kathleen Freeman, *The Pre-Socratic Philosophers* (Oxford: Basil Blackwell, 1949), DK68, 117.

CHAPTER THREE

1 Niels Bohr, *Studier over metallernes elektrontheori* (Copenhagen, 1911), p. 5; quoted by John L. Heilbron and Thomas S. Kuhn, "The Genesis of the Bohr Atom", in *Historical Studies in the Physical Sciences, Vol. 1,* ed. by Russell McCormmach (Philadelphia: University of Pennsylvania Press, 1969), p. 215.
2 *Idem.*
3 *Idem.*
4 *NBA*, Lecture Notes for "Amherst Lectures", October 12–22, 1923; "Science Club Lecture", October 29, 1923; and "Silliman Lecture", 1923. This outlook is also particularly marked in Bohr's first truly "philosophical" essay on atomic physics, from a lecture delivered to the Sixth Scadinavian Mathematical Congress on 30 August, 1925, "Atomic Theory and Mechanics", *Nature 116*, 845, reprinted in *Atomic Theory and the Description of Nature*, where he comments, "there exists a fundamental difference between an atom and a planetary system. The atom must have a *stability* which presents features quite foreign to mechanical theory. Thus, the mechanical laws permit a continuous variation of the possible motions, which is entirely at variance with the definiteness of the properties of the elements." pp. 30–31.
5 *Cf.*, Thomas S. Kuhn, *Black-Body Theory and the Quantum Discontinuity 1894–1912* (New York and Oxford: Clarendon Press and Oxford University Press, 1978).
6 *NBA*, Bohr Scientific Correspondence (hereafter designated "*BSC*"); letter from Niels Bohr to C.W. Oseen, 1 Dec. 1911.
7 *NBA:BSC*, letter from Niels Bohr to Harald Bohr, 12 Jun. 1912.
8 *NBA:BSC*, letter from Niels Bohr to Harald Bohr, 19 Jun. 1912; italics mine.
9 For example, Feuer, *Generations of Science*, pp. 121–126, 131–146.
10 *NBA:BSC*, letter from Niels Bohr to C.W. Oseen, 29 Jan. 1926. *Cf.* also, Jammer, *Conceptual Development*, pp. 87–88, 110; and Niels Bohr, *The Theory of Spectra and Atomic Constitution* (Cambridge: The University Press, 1922), pp. 12–13.
11 The principle was first used implicitly in Bohr's 1913 papers, but was not stated explicitly until 1920, in Bohr, *The Theory of Spectra*, pp. 23–24. For further discussion of the correspondence principle *cf.*, Jammer, *Conceptual Development*, pp. 109–118. A philosophical analysis of the role of this principle in the development of complementarity is given by Aage Petersen, in "On the Philosophical Significance of the Correspondence Argument", in *Boston Studies in the Philosophy of Science, Vol. V*, ed. by R.S. Cohen and M. Wartofsky (Dordrecht: D. Reidel, 1969), pp. 242–252.
12 Bohr, *Atomic Theory and the Description of Nature*, p. 92.
13 *AHQP*, Interview with Oskar Klein, conducted by T.S. Kuhn and J.L. Heilbron (Copenhagen, 15 Jul. 1963), Tape 83a, transcript p. 3.
14 *AHQP*, "Last Interview", p. 4.
15 *AHQP*, Interview with Werner Heisenberg, conducted by T.S. Kuhn (Munich, 15 Feb. 1963), Tape 49a, p. 15.

16 *NBA:BSC*, letter from Niels Bohr to C.W. Oseen, 28 Sep. 1914.
17 N. Bohr, H.A. Kramers, and J.C. Slater, "The Quantum Theory of Radiation", *Philosophical Magazine 47* (1924), 785; reprinted in *Sources of Quantum Mechanics*, ed. by B.L. van der Waerden (Amsterdam: North-Holland Publishing Co., 1967), p. 159.
18 *Ibid.*, 786, p. 160; italics mine.
19 *Ibid.*, 787, p. 161; italics mine.
20 *NBA:BSC*, unsent letter from Niels Bohr to C.G. Darwin, *circa* 1919.
21 *AHQP*, Interview with Klein, Tape 83a, pp. 8–9.
22 *NBA:BSC*, letter from Niels Bohr to A.A. Michelson, 7 Feb. 1924; italics mine.
23 *NBA:BSC*, letter from Niels Bohr to R.H. Fowler, 5 Dec. 1924.
24 *NBA:BSC*, letter from Niels Bohr to R.H. Fowler, 21 Apr. 1925; italics mine.
25 Werner Heisenberg, "Quantum Theory and Its Interpretation", in Rozental (ed.), *Niels Bohr*, pp. 99–100.
26 *Idem.*
27 *Ibid.*, p. 98.
28 *AHQP*, Interview with Werner Heisenberg, conducted by T.S. Kuhn (Munich, 13 Feb. 1963), Tape 48b, p. 13.
29 *AHQP*, Interview with Werner Heisenberg, conducted by T.S. Kuhn (Munich, 25 Feb. 1963), Tape 52a, p. 6.
30 *Idem.*
31 *AHQP*, Interview with Werner Heisenberg, conducted by T.S. Kuhn (Munich, 11 Feb. 1963), Tape 48a, p. 13.
32 *AHQP*, Interview with Werner Heisenberg, conducted by T.S. Kuhn (Munich, 19 Feb. 1963), Tape 50b, p. 22.
33 *AHQP*, Interview with Werner Heisenberg, conducted by T.S. Kuhn (Munich, 28 Feb. 1963), Tapes 50a–b, p. 13.
34 *AHQP*, Heisenberg interview, Tape 49a, pp. 17–18.
35 *AHQP*, Heisenberg interview, Tape 52a, pp. 12–13.
36 *NBA:BSC*, letter from Niels Bohr to Wolfgang Pauli, 11 Dec. 1924.
37 *AHQP*, Interview with Werner Heisenberg, conducted by T.S. Kuhn and J.L. Heilbron (Copenhagen, 5 Jul. 1963), Tape 74a, p. 11.
38 Max Jammer, *The Philosophy of Quantum Mechanics*, p. 125.
39 Niels Bohr, "Über die Wirkung von Atomen bei Stössen", *Zeitschrift für Physik 34* (1925), p. 142. *Cf.* Martin J. Klein, "The First Phase of the Bohr–Einstein Dialogue", *Historical Studies in the Physical Sciences*, *Vol. 2*, ed. by Russell McCormmach (Philadelphia: University of Pennsylvania Press, 1970), pp. 33–36.
40 *NBA:BSC*, letter from Niels Bohr to John Slater, 28 Jan. 1926.
41 *NBA:BSC*, letter from Niels Bohr to C.W. Oseen, 29 Jan. 1926.
42 *AHQP*; Heisenberg interviews, Tape 39a, p. 15 and Tape 52a, p. 10.
43 *NBA:BSC*, letter from Niels Bohr to R.H. Fowler, 26 Oct. 1926.
44 *NBA:BSC*, letter from Niels Bohr to C.G. Darwin, 24 Nov. 1926.
45 *AHQP*, Heisenberg interview, Tape 52a, p. 11.
46 *AHQP*, Heisenberg interview, Tape 49a, p. 19.
47 *AHQP*, Heisenberg interview, Tape 74a, p. 10.
48 *AHQP*, Heisenberg interview, Tape 52a, pp. 16–17.
49 *Idem.*

50 *AHQP*, Heisenberg interview, Tape 74a, pp. 13–14.
51 *AHQP*, Heisenberg interview, Tape 52a, p. 15.
52 *AHQP*, Interview with Werner Heisenberg, conducted by T.S. Kuhn (Munich, 27 Feb. 1963), Tape 52b, p. 26; italics mine.
53 *NBA:BSC*, letter from Niels Bohr to Albert Einstein, 13 Apr. 1927.
54 Niels Bohr, "Atomic Theory and Mechanics" (1925), in *Atomic Theory and the Description of Nature*, p. 25.
55 *Ibid.*, p. 26.
56 *Ibid.*, pp. 34–35.
57 *Ibid.*, p. 36.
58 *Ibid.*, p. 39.
59 *Ibid.*, p. 48.
60 *Ibid.*, p. 49.
61 *Ibid.*, p. 50.

CHAPTER FOUR

1 Bohr, "The Quantum Postulate and the Recent Development of Atomic Theory" (hereafter cited as "Como Lecture"), first published in *Supplement to Nature 121* (1928), 580–590; reprinted in *Atomic Theory and the Description of Nature*, pp. 54–55. A substantially different version appears in *Atti del Congresso Internazionale dei Fisici* (Bologna, 1928), pp. 565–588.
2 *Ibid.*, p. 52.
3 *Idem.*
4 Jammer, *The Philosophy of Quantum Mechanics*, p. 122.
5 *NBA*, Notes and Manuscripts for Como Lecture, 1927.
6 *AHQP*, Interview with Oskar Klein conducted by J.L. Heilbron and L. Rosenfeld (Copenhagen, 28 Feb. 1963), Tape 59a, pp. 9–10.
7 *AHQP*, Klein interview, Tape 83a, p. 22.
8 *AHQP*, Klein interview, Tape 59a, pp. 10–11.
9 *NBA:BSC*, letter from Niels Bohr to Werner Heisenberg, 24 Aug. 1927.
10 *NBA:BSC*, letter from Niels Bohr to C.G. Darwin, 29 Aug 1927; italics mine.
11 *AHQP*, Klein interview, Tape 59a, p. 11.
12 *NBA*, Notes and manuscripts for Como Lecture, 1927. Bohr sent this typescript to Darwin, dated October 12–13, 1927, twenty-six days *after* the Como lecture.
13 Niels Bohr, "Como Lecture", in *Atomic Theory and the Description of Nature*, p. 53.
14 *Ibid.*, pp. 53–54.
15 *Cf.* Section 3 of Chapter Five below.
16 *Ibid.*, p. 54.
17 *Cf.* Section 4 of Chapter Five below.
18 *Idem.*
19 *Ibid.*, pp. 54–55.
20 *Ibid.*, p. 55.
21 *Idem.*
22 *Ibid.*, pp. 55–56.
23 *Ibid.*, p. 56.
24 *Ibid.*, pp. 56–57.
25 Niels Bohr, "The Atomic Theory and the Fundamental Principles Underlying the De-

scription of Nature", elaboration of a lecture delivered to 18th Annual Meeting of Scandinavian Natural Scientists, August 1929, reprinted in *Atomic Theory and the Description of Nature*, pp. 108–109.

26 This will be discussed further in Section 3 of Chapter Five.

27 Bohr, "Causality and Complementarity", *Phil. Sci. 4*, 294.

28 Bohr, "Como Lecture", *Atomic Theory and the Description of Nature*, p. 56.

29 *Idem*; italics mine.

30 Niels Bohr, "The Quantum of Action and the Description of Nature", contributed to "a jubilee pamphlet" in honor of Max Planck; originally published in *Naturwissenschaften 17* (1929), 483–486; reprinted in *Atomic Theory and the Description of Nature*, p. 98.

31 Bohr, "Causality and Complementarity", *Phil. Sci. 4*, 293–294.

32 Niels Bohr, "The Causality Problem in Atomic Physics", a lecture given in Warsaw, May 1939, published in *New Theories in Physics* (Paris: International Institute for Intellectual Cooperation, 1939), p. 17.

33 Bohr, "Newton's Principles", *Newton Tercentenary Celebrations*, p. 58.

34 *Ibid.*, p. 59

35 Bohr, "Causality and Complementarity", *Phil. Sci. 4*, 293.

36 Bohr, "The Atomic Theory and the Fundamental Principles Underlying the Description of Nature", *Atomic Theory and the Description of Nature*, p. 114.

37 Niels Bohr, "Atomic Stability and Conservation Laws", in *Atti del Convegno di Fisica Nucleare della "Fondazione Alessandro Volta"* (Rome: Reale Academia d'Italia, 1932), p. 6.

38 Niels Bohr, "Chemistry and the Quantum Theory of Atomic Constitution" (Bohr's Faraday Lecture of 8 May 1930), published in *Journal of the Chemical Society* (1932), 376.

39 Niels Bohr, "The Atomic Theory and the Fundamental Principles Underlying the Description of Nature", *Atomic Theory and the Description of Nature*, p. 106.

40 Bohr, "Causality and Complementarity", *Phil. Sci. 4*, 294.

41 Heisenberg, "Quantum Theory and Its Interpretation", in Rozental (ed.) *Niels Bohr*, p. 105.

42 Werner Heisenberg, *The Physical Principles of the Quantum Theory* (Chicago: University of Chicago Press, 1930), p. 15.

43 *Ibid.*, p. 14.

44 *NBA*, Bohr MSS "*Statistik und Reciprocität*", 1929. Bohr alludes to this possibility in "The Quantum of Action and the Description of Nature", *Atomic Theory and the Description of Nature*, p. 95.

45 *NBA:BSC*, letter from Bertrand Russell to Niels Bohr, 30 Sep. 1935.

46 Bohr, "Como Lecture", *Atomic Theory and the Description of Nature*, p. 68.

47 *Ibid.*, p. 55; italics mine.

CHAPTER FIVE

1 Niels Bohr, "Can the Quantum Mechanical Description of Reality be Considered Complete?", *Physical Review 48* (1935), 702.

2 *Cf.* the excellent analysis of this debate in Hooker, "The Nature of Quantum Mechanical Reality", in Colodny (ed.), *Paradigms and Paradoxes*, pp. 67–302.

3 Niels Bohr, "Discussions with Einstein", in Schilpp (ed.), *Albert Einstein*, pp. 199–242.

4 *AHQP*, Bohr, "Last Interview", p. 4.

5 Jammer, *The Philosophy of Quantum Mechanics*, p. 187, quoting A. Einstein, *Lettres à Maurice Solovine* (Paris: Gauthier-Villars, 1956), p. 74.
6 *Ibid.*, quoting Philipp Frank, *Einstein: His Life and Times* (New York: Knopf, 1947), p. 216.
7 For an historical analysis cf. *Ibid.*, pp. 121–211; for a conceptual analysis *cf.* Hooker, *Paradigms and Paradoxes*, pp. 67–302.
8 A. Einstein, B. Podolsky, and N. Rosen, "Can Quantum Mechanical Description of Reality Be Considered Complete?", *Physical Review 47* (1935), 777.
9 *Idem.*
10 Niels Bohr, "Quantum Mechanics and Physical Reality", *Nature 136* (13 Jul. 1935), 65.
11 A. Einstein, B. Podolsky, and N. Rosen, "Can Quantum Mechanical Description of Reality Be Considered Complete?", *Physical Review 47* (1935), 780.
12 Niels Bohr, "Can the Quantum Mechanical Description of Reality be Considered Complete?", *Physical Review 48* (1935), 699.
13 *Ibid.*, 700.
14 Bohr, "The Causality Problem in Atomic Physics", *New Theories in Physics*, pp. 19–20.
15 *Ibid.*, p. 21.
16 Jammer, *The Philosophy of Quantum Mechanics*, p. 127.
17 Niels Bohr, "Discussions with Einstein", in Schilpp (ed.), *Albert Einstein*, p. 235.
18 *Cf.* Jammer, *The Philosophy of Quantum Mechanics*, for a discussion of Einstein's views towards so-called "hidden-variable theories".
19 Bohr, "Last Interview", p. 7.
20 Bohr, "Chemistry and the Quantum Theory of Atomic Constitution", *J. Chem. Soc.* (1932), 375.
21 Bohr, "Causality and Complementarity", *Phil. Sci. 4*, 292–293.
22 Niels Bohr, "Introductory Survey" to *Atomic Theory and the Description of Nature*, pp. 11–12.
23 Bohr, "Biology and Atomic Physics" (lecture given to the Galvani Congress in Bologna, Oct. 1937), in *Atomic Physics and Human Knowledge*, p. 19.
24 Bohr, "The Causality Problem in Atomic Physics", *New Theories in Physics*, p. 24.
25 *Ibid.*, p. 22.
26 Bohr, "Newton's Principles and Modern Atomic Mechanics", *Newton Tercentenary Celebrations*, pp. 59–60.
27 Even as early as 1927 in the original Como lecture, in addition to speaking of complementarity between two modes of description, Bohr spoke of the complementarity of wave and particle pictures; *Atomic Theory and the Description of Nature*, p. 56. In his reply to the EPR paper, Bohr speaks of "complementary physical quantities" of position and momentum; "Can Quantum Mechanical Description of Physical Reality be Considered Complete?", *Physical Review 48* (1935), 700. In 1948 he uses the expression "complementary evidence" to refer to different observations obtained in different experiments; "On the Notions of Causality and Complementarity", *Dialectica 2* (1948), 314. In the Compton Lectures of 1957 he speaks of "complementary information" and "complementary experiences"; *NBA*, transcript of Karl Taylor Compton Lectures, Sixth Lecture, delivered 26 Nov. 1957, pp. 1 and 3. However, in his later lectures most commonly he refers to "complementary phenomena"; *e.g.*, *Atomic Physics and Human Knowledge*, pp. 90, 99.
28 In 1955 Heisenberg's former student and associate, C.F. von Weizsäcker suggested that

Bohr had unintentionally developed two notions of complementarity: "parallel complementarity" is a relation between different descriptive concepts such as position and momentum, while "circular complementarity" holds between the two modes of description; "Komplementarität und Logik", *Naturwissenschaften 42* (1955), 521–529, 545–555. When von Weizsäcker presented this idea to Bohr, the latter categorically denied any such equivocal use of "complementary". However, Bohr's reply to von Weizsäcker is basically a rejection of the latter's attempt to use complementarity as the rationale for developing a new logic (a "quantum logic") to describe the character of the reality represented in quantum theory. Bohr saw such an approach (rightly or wrongly) as undercutting his insistence on describing the phenomena in classical terms; thus Bohr's conviction that von Weizsäcker had misunderstood complementarity was based on grounds other than the distinction between parallel and circular complementarity; *NBA: BSC*, letter from von Weizsäcker to Bohr, 17 Jan. 1956; letter from Bohr to von Weizsäcker 5 Mar. 1956. Although I now believe, as I have argued in the text, that Bohr intended a singular meaning for "complementary", elsewhere I have distinguished two different senses of "complementary"; *cf.*, my "Complementarity and the Description of Experience", *International Philosophical Quarterly 27* (1977), 389. At the time this article was written (August 1974), I was not aware of von Weizsäcker's distinction.

29 *NBA*, Bohr MSS, Notes for Como Lecture, 1927.

30 Bohr, "Como Lecture", *Atomic Theory and the Description of Nature*, p. 56.

31 Bohr, "Biology and Atomic Physics", *Atomic Physics and Human Knowledge*, p. 16.

32 Bohr, "The Causality Problem in Atomic Physics", *New Theories in Physics*, pp. 24–25.

33 Bohr, "Natural Philosophy and Human Cultures", *Nature 143* (1939), 268, reprinted in *Atomic Physics and Human Knowledge*, p. 26.

34 Bohr, "Newton's Principles and Modern Atomic Mechanics", *Newton Tercentenary Celebrations*, p. 60.

35 Bohr, "On the Notions of Causality and Complementarity", *Dialectica 2* (1948), p. 314.

36 *NBA*, Notes and Manuscripts for Como Lecture, 1927.

37 Niels Bohr, "Discussions with Einstein", in Schilpp (ed.), *Albert Einstein*, p. 202.

38 Bohr, "On the Notions of Causality and Complementarity", *Dialectica 2* (1948), p. 317; italics mine.

CHAPTER SIX

1 Jammer, *The Philosophy of Quantum Mechanics*, p. 247.

2 Bohr, "Unity of Knowledge—1954", in *Atomic Physics and Human Knowledge*, pp. 67–68.

3 Bohr, "On the Notions of Causality and Complementarity", *Dialectica 2*, 318.

4 An analogy of this sort is suggested by Arthur Fine in "Some Conceptual Problems of Quantum Theory", in Colodny (ed.), *Paradigms and Paradoxes*, p. 26.

5 Bohr, "Causality and Complementarity", *Phil. Sci. 4*, 295.

6 Bohr, "Last Interview", p. 1.

7 *Ibid.*, p. 2.

8 Bohr, "Como Lecture", *Atomic Theory and the Description of Nature*, p. 91.

9 Bohr, "The Quantum of Action and the Description of Nature", *Atomic Theory and the Description of Nature*, p. 96.

10 Bohr, "Physical Science and the Study of Religions", in *Studia Orientalia Ioanni Petersen*, pp. 389–390.

11 Bohr, "The Quantum of Action and the Description of Nature", *Atomic Theory and the Description of Nature*, p. 100.
12 *Idem.*
13 Bohr, "The Atomic Theory and the Fundamental Principles Underlying the Description of Nature", in *Atomic Theory and the Description of Nature*, pp. 116–117.
14 Niels Bohr, "Light and Life", Nature 131 (1933), 457. A somewhat different version appears in *Atomic Physics and Human Knowledge*, pp. 3–12.
15 Bohr, "Introductory Survey" to *Atomic Theory and the Description of Nature*, p 4.
16 *Ibid.*, p. 3.
17 Bohr, "Light and Life", *Nature 131* (1933), 421 and 457.
18 *Ibid.*, 422.
19 *Idem.*
20 *Idem.*
21 *Idem.*
22 Bohr, "Biology and Atomic Physics", *Atomic Physics and Human Knowledge*, p. 20.
23 *Ibid.*, pp 20–21.
24 Bohr, "Physical Science and the Problem of Life", *Atomic Physics and Human Knowledge*, p. 97.
25 *Ibid.*, p. 100; italics mine.
26 *Ibid.*, p. 102.
27 Bohr, "The Connection Between The Sciences—1960", *Essays 1958–1962*, p. 21.
28 Bohr, "Light and Life Revisited", *Essays 1958–1962*, p. 26.
29 *Idem.*
30 *Idem.*
31 *Idem.*

CHAPTER SEVEN

1 Bohr, "Last Interview", p. 3.
2 *Idem.*
3 *Cf.*, C.A. Hooker, "The Nature of Quantum Mechanical Reality", in Colodny (ed.), *Paradigms and Paradoxes*, pp. 193–209.
4 Bohr, "Introductory Survey" to *Atomic Theory and the Description of Nature*, p. 1
5 Bohr, "Unity of Knowledge—1954", in *Atomic Physics and Human Knowledge*, p. 80.
6 Bohr, "Introductory Survey" to *Atomic Theory and the Description of Nature*, p. 1
7 Bohr, "Quantum Physics and Philosophy", in *Essays 1958–1962*, p. 7.
8 Bohr, "Unity of Knowledge—1954", in *Atomic Physics and Human Knowledge*, p. 91; italics mine.
9 *Idem*; Bohr's use of "consequent" in translating the Danish "*konsekvent*" should more accurately be translated "consistent".
10 Bohr, "Quantum Physics and Philosophy", in *Essays 1958–1962*, p. 7.
11 *Ibid.*, p. 3.
12 Although Bohr's position that "subject" and "object" are distinguished only in the *description* of experience may have been influenced by William James' radical empiricism, there is no direct evidence of any such connection. Furthermore, the apparent similarity to Russell's

version of "neutral monism" or Carnap's *Aufbau* program is somewhat deceiving. These theories were essentially logical reconstructionist analyses of the mind/object relation. Bohr's position does not address this issue at all, but is concerned with the consequences of an empirical fact—that an "observation" (in atomic physics) involves an "uncontrollable" interaction—on the objectivity of empirical science.

13 Bohr, "Unity of Knowledge—1954", in *Atomic Physics and Human Knowledge*, pp. 67–68.

14 Bohr, "The Quantum of Action and the Description of Nature", *Atomic Theory and the Description of Nature*, p. 95.

15 Bohr, "The Causality Problem in Atomic Physics", *New Theories in Physics*, p. 24.

16 Bohr, "Unity of Knowledge—1954", in *Atomic Physics and Human Knowledge*, p. 74; italics mine.

17 *NBA:BSC*, letter from Wolfgang Pauli to Niels Bohr, 15 Feb. 1955.

18 *NBA:BSC*, letter from Niels Bohr to Wolfgang Pauli, 2 Mar. 1955.

19 *NBA:BSC*, letter from Wolfgang Pauli to Niels Bohr, 11 Mar. 1955.

20 *NBA:BSC*, letter from Niels Bohr to Wolfgang Pauli, 25 Mar. 1955.

21 *Cf.* C.A. Hooker, "The Nature of Quantum Mechanical Reality", in Colodny (ed.), *Paradigms and Paradoxes*, pp. 135–138, 156, and 168–172; Meyer-Abich, *Korrespondenz, Individualität, und Komplementarität*, pp. 140, 176–178; C.F. von Weizsäcker, *The World View of Physics* (London: Routledge and Kegan Paul, 1952), pp. 120–124; Aage Petersen, *Quantum Physics and the Philosophical Tradition* (Cambridge, Mass.: M.I.T. Press, 1968), p. 142; and P.K. Feyerabend, "On a Recent Critique of Complementarity, Part II", *Philosophy of Science 36* (1969), pp. 82 and 92–93, and "Problems of Microphysics", in Colodny (ed.), *Frontiers of Science and Philosophy*, pp. 229–230.

22 Bohr, "The Unity of Human Knowledge—1960", *Essays 1958–1962*, pp. 9–10.

23 Bohr, "Como Lecture", *Atomic Theory and the Description of Nature*, p. 54.

CHAPTER EIGHT

1 Bohr, "The Quantum of Action and the Description of Nature", in *Atomic Theory and the Description of Nature*, p. 93.

2 *NBA:BSC*, letter from Niels Bohr to Harald Bohr, 19 Jun. 1912; italics mine.

3 *NBA*, Lecture notes for Amherst Lectures, Science Club Lecture, and Silliman Lecture, 1923.

4 Bohr, "The Atomic Theory and the Fundamental Principles underlying the Description of Nature", in *Atomic Theory and the Description of Nature*, pp. 102–103; italics mine.

5 *Cf. AHQP*, Interview with Philipp Frank, conducted by T.S. Kuhn (Cambridge, Mass., 16 Jul. 1962), Tape 25a, transcript pp. 5–6.

6 Heisenberg, "Quantum Theory and Its Interpretation", in Rozental (ed.), *Niels Bohr*, p. 98.

7 Heisenberg, *Physics and Beyond*, p. 61.

8 *AHQP*, Interview with Werner Heisenberg, conducted by T.S. Kuhn (Munich, 30 Nov. 1962), Tape 39a, pp. 3–4.

9 Bohr, "Introductory Survey" to *Atomic Theory and the Description of Nature*, p. 18.

10 *AHQP*, Interview with Werner Heisenberg, conducted by T.S. Kuhn (Munich, 25 Feb. 1963), Tape 52a, p. 6.

11 Bohr, "Last Interview", p. 3.

12 *Idem.*

13 *AHQP*, Interview with Werner Heisenberg, conducted by T.S. Kuhn (Munich, 13 Feb. 1963), Tape 48b, p. 7.

14 *Cf.* Adolf Grunbaum, "Complementarity in Quantum Physics and Its Philosophical Generalization", *Journal of Philosophy 34* (1957), 713–727.

15 Bohr, "Introductory Survey", to *Atomic Theory and the Description of Nature*, pp. 1–2.

16 *NBA*, Bohr MSS, Compton Lectures, Sixth Lecture, 26 Nov. 1957, transcript p. 1.

17 Bohr, "Natural Philosophy and Human Cultures", *Atomic Physics and Human Knowledge*, p. 26.

18 *NBA*, Bohr MSS, Compton Lectures, Sixth Lecture, pp. 2–3.

19 Bohr, "Physical Science and the Problem of Life", in *Atomic Physics and Human Knowledge*, p. 99; italics mine.

20 Bohr, "Quantum Physics and Philosophy", *Essays 1958–1962*, p. 5.

21 Max Born, "Symbol and Reality", in *Natural Philosophy of Cause and Chance* (New York: Dover Publications, 1964), pp. 227–232, *passim.*

22 *NBA:BSC*, letter from Niels Bohr to Max Born, 2 Mar. 1953; Bohr quotes Born from a preprint Born had sent to him of his article, "The Interpretation of Quantum Mechanics", reprinted in Max Born, *Physics in My Generation* (London: Pergamon Press, 1956), p. 144.

23 *NBA:BSC*, letter from Niels Bohr to Max Born, 2 Mar. 1953.

24 *NBA:BSC*, letter from Max Born to Niels Bohr, 10 Mar. 1953.

25 *NBA:BSC*, letter from Niels Bohr to Max Born, 26 Mar. 1953.

26 Bohr, "Unity of Human Knowledge—1954", in *Atomic Physics and Human Knowledge*, pp. 78–79.

27 Aage Petersen, "The Philosophy of Niels Bohr", *Bulletin of the Atomic Scientists 9* (1963), p. 8.

28 Bohr, "Causality and Complementarity", *Phil. Sci. 4*, 294.

29 Bohr, "Unity of Human Knowledge—1954", in *Atomic Physics and Human Knowledge*, p. 82.

INDEX